Your passport to a wild Journey!

An Adventure in
Tropical & Temperate
Rainforests

5 Continents 7 Rainforests
Hundreds of extraordinary animals & plants!

By Deanna Holm

An Adventure in Tropical & Temperate
Rainforests by Deanna Holm
www.discoverunitstudies.com

Published by Kitsap Publishing
In cooperation with Discover Unit Studies

Edited by Cheryl Stickney

ISBN 978-0-997899-80-1

Library of Congress Control Number:
2017949233

First edition, 2017

Front cover photo credits: background forest
by Galyna Andrushko @123rf.com, sloth by Eric
Isselee @123rf.com, chameleon by Vera
Kuttelvaserova Stuchelova @123rf.com,
passion fruit by Madlen @123rf.com, rafflesia
by Somsak 2503 @shutterstock.com

Table of Contents

Ch. 1 All About Rainforests

Ch. 2 Tropical Rainforests

Ch. 3 Temperate Rainforests

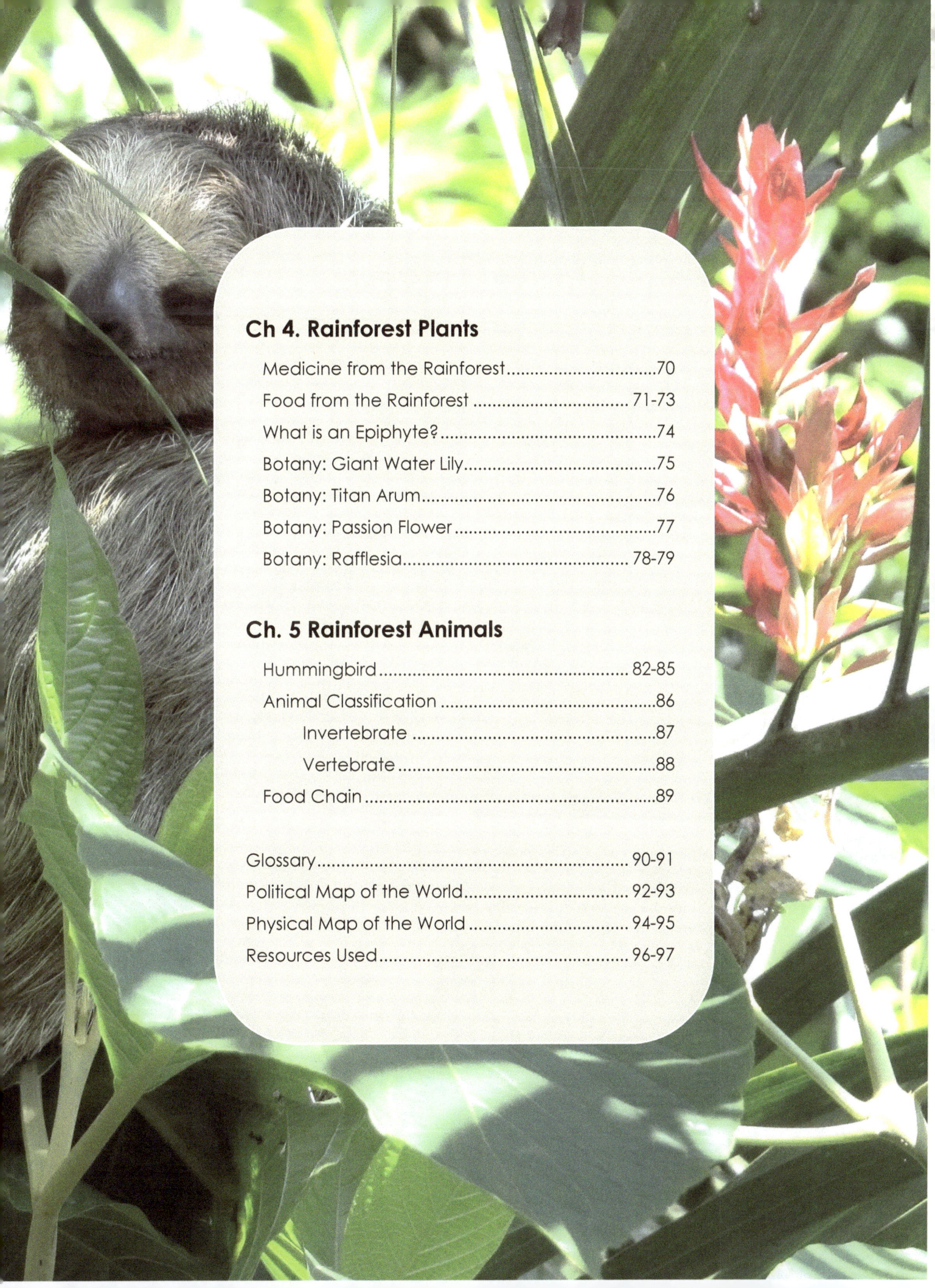

Ch 4. Rainforest Plants

Ch. 5 Rainforest Animals

Chapter

1

All About Rainforests

Rainforests of the World

Rainforests are filled with fascinating creatures and bizarre plants. They are scattered on every continent except Antarctica. Let's go on an adventure to find these rainforests and discover what you would find there. As we travel the world we will learn about two kinds of rainforests. We will also learn many interesting facts about the animals, plants, and resources of 7 major rainforests. As you journey with me in this book you will gain a greater understanding and appreciation of this amazing world God has created for us.

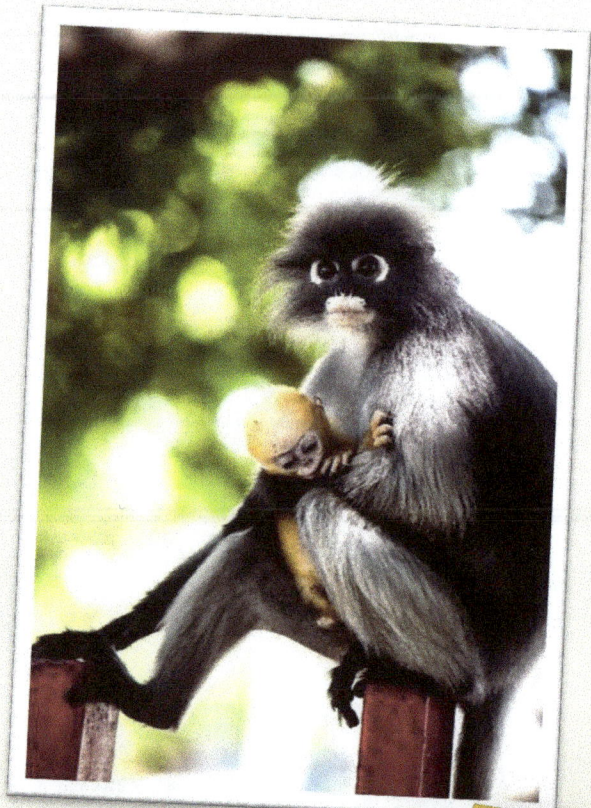

FUN FACT!

The Spectacle Leaf Monkey gets its name because it has white circles around its eyes which look like glasses or "spectacles". Their babies are born a bright orange color and turn gray by the time they are 10 months old.

Bananas

Poison Dart Frog

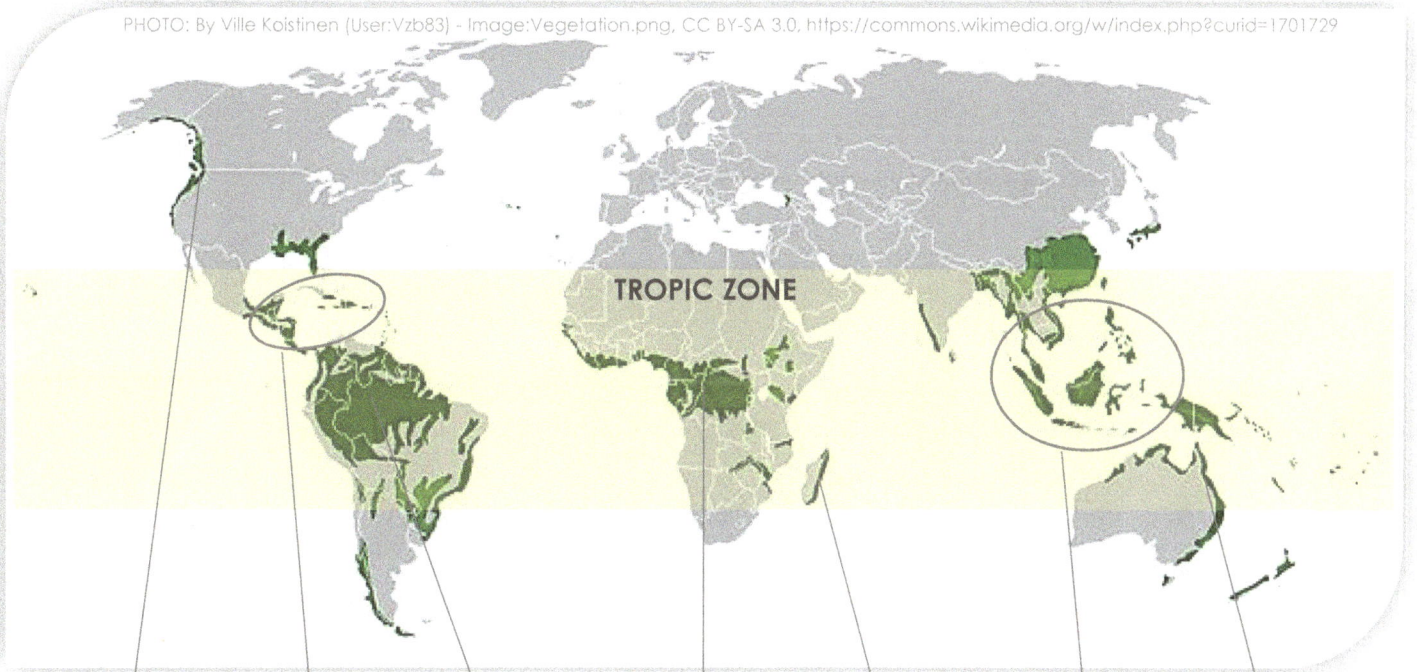

TROPIC ZONE

| THE HOH RAINFOREST (temperate) | CENTRAL AMERICA RAINFORESTS | AMAZON RAINFOREST | CONGO RAINFOREST | MADAGASCAR RAINFOREST | SOUTHEAST ASIA RAINFORESTS | AUSTRALIA RAINFOREST |

Orchid

Golden Lion Tamarin

Chimpanzee

Amazon River

9

A Trail leading into a Rainforest in Thailand

What is a Rainforest?

A rainforest is a dense green forest with tall trees. It has lots of rain with no dry season. These forests receive anywhere from one hundred to four hundred inches of rain every year.

Continual rainfall allows plants to grow large. It also provides food and water for a wide variety of animals. This creates an **ecosystem** that has captured the imagination of young and old.

Two Kinds of Rainforests

There are two major kinds of rainforests, tropical and temperate. The main difference between the two is the temperature. One has changing seasons and the other is hot all the time. Both receive lots of regular rainfall and are very green and alive with a variety of plants and animals.

Tropical rainforests are hot all the time. They lie in the tropic zone near the **equator** line.

Temperate rainforests have changing seasons from summer to winter. These forests are found above or below the tropic zone.

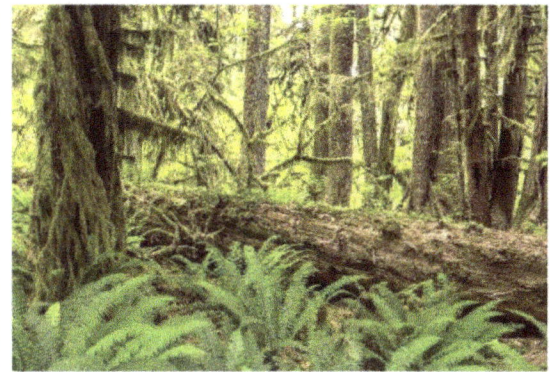

ecosystem (ē-kō-ˌsi-stəm) the whole group of living and nonliving things that make up an environment and affect each other

Imaginary Lines

Imaginary lines on a map help us to locate a specific place. These lines are called latitude and longitude. They are only found on maps and globes. Latitude lines run East to West and longitude lines run from the North Pole to the South Pole. The Tropic of Cancer, the Equator, and the Tropic of Capricorn are latitude lines. The area between these lines is known as the tropics. Here it is warm all year long because it is closest to the sun. This is where all tropical rainforests are found. The earth sits on an imaginary axis that is tilted which gives Earth its seasons as it moves around the sun.

Diagram 1

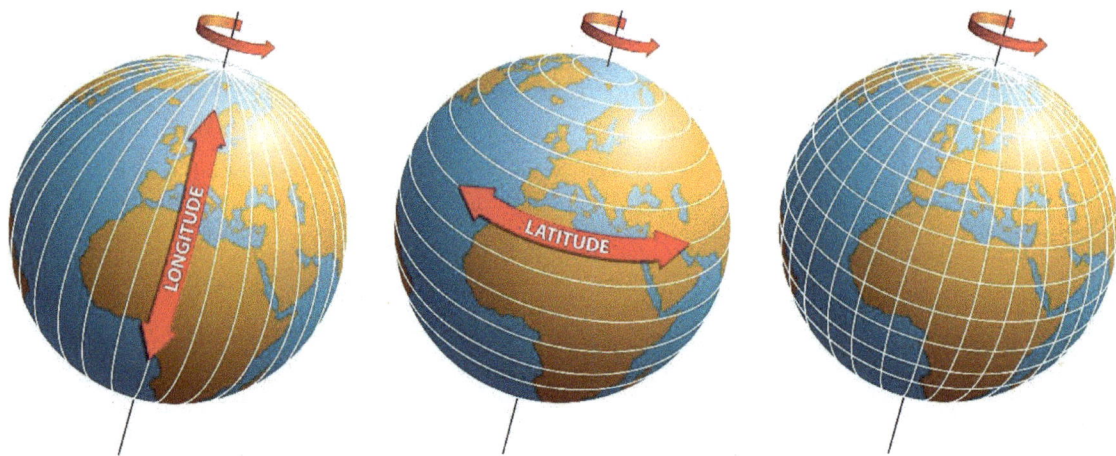

LONGITUDE

LATITUDE

Diagram 2

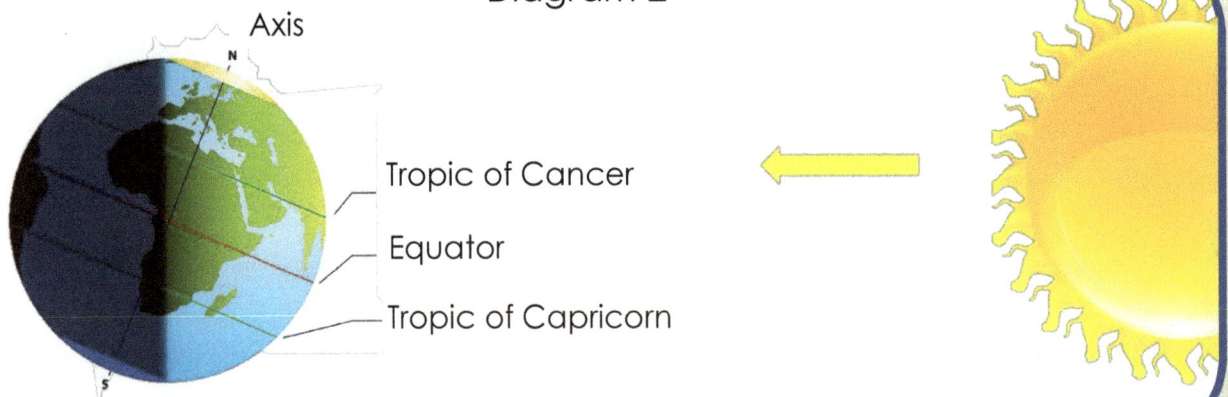

Axis

N

Tropic of Cancer

Equator

Tropic of Capricorn

S

Layers of the Rainforest

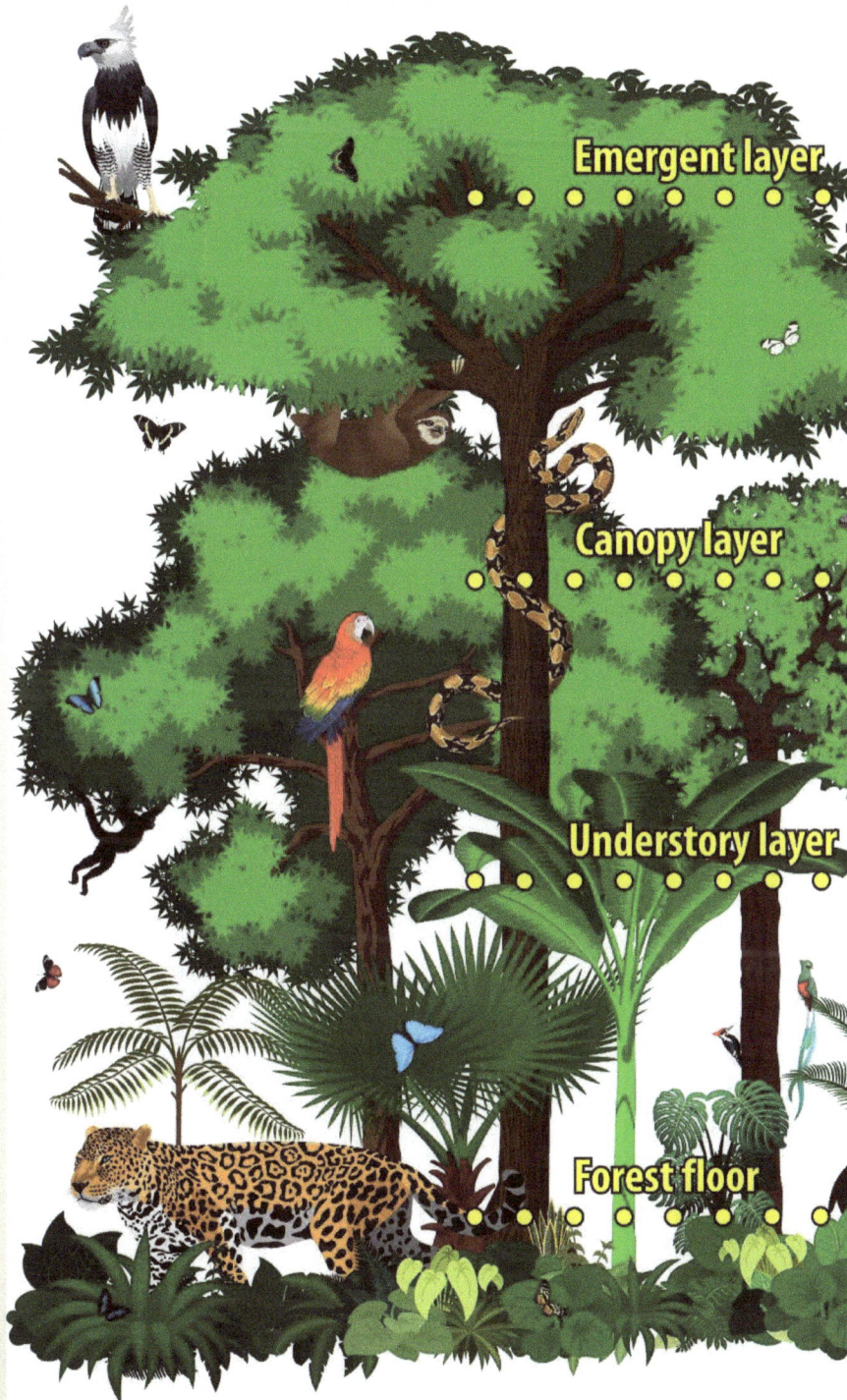

Emergent layer

Canopy layer

Understory layer

Forest floor

Emergent Layer

In this layer, giant trees stick out above the canopy. They are much taller than the canopy trees. Many birds and insects live here.

Canopy Layer

This layer forms a cover over the lower layers and is full of life. It is home to many insects, birds, reptiles, and mammals.

Understory Layer

This layer is cool and dark. It is between the canopy and the ground.

Forest Floor

The ground layer of the rainforest is teeming with insect life and is home to the biggest animals of the rainforest.

The Water Cycle

The water we drink today is the same water that has been around since creation. This is because water is recycled through a process called the water cycle. Water which falls from the clouds is called **precipitation**. This is the name for rain, snow, or hail. **Gravity** pulls the water downward. The water falls to the ground and forms rivers or soaks into the ground and makes its way into lakes or the ocean. As the sun shines down on Earth it begins to warm up the water turning it into a gas called water vapor. The water evaporates into the sky. The vapor accumulates and forms clouds. When the clouds become heavy it begins to precipitate again completing the continuous water cycle.

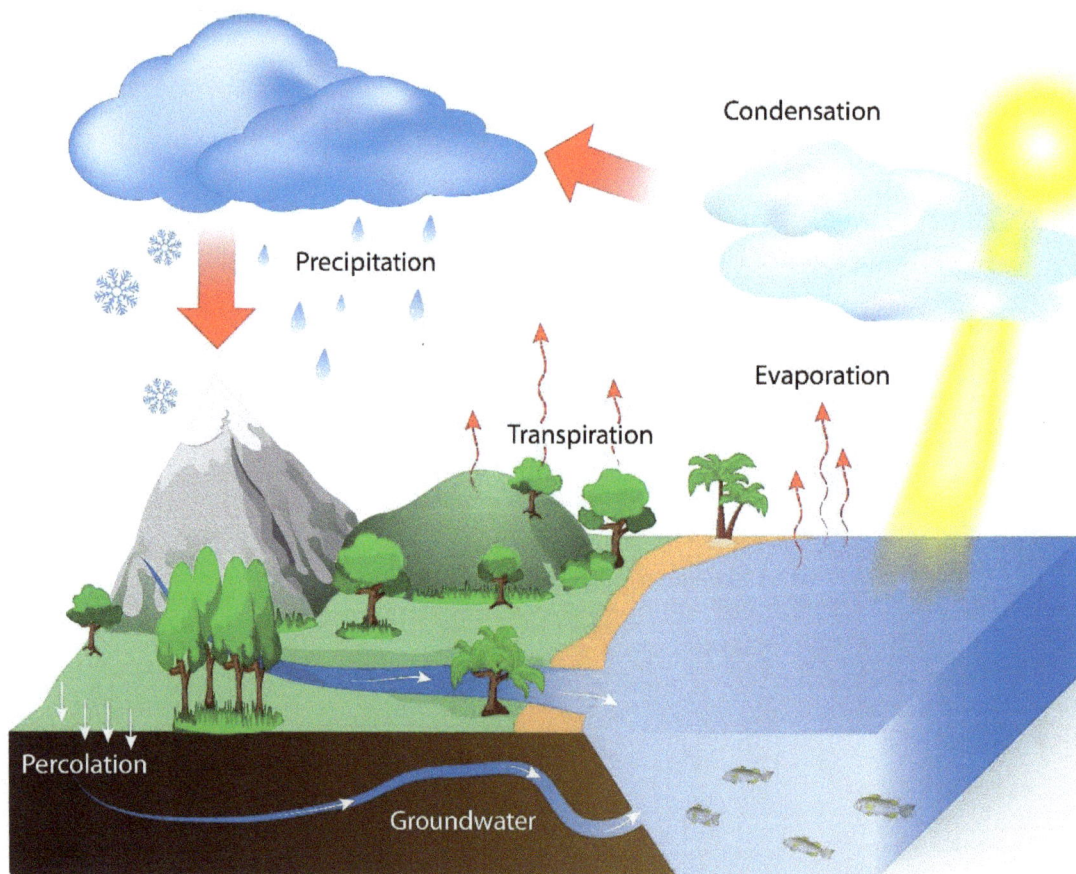

Condensation

Precipitation

Evaporation

Transpiration

Percolation

Groundwater

gravity (ˈgra-və-tē) the natural force that tends to cause physical things to move towards each other: the force that causes things to fall towards the Earth

Photosynthesis

Plants use energy from the sun to turn water and carbon dioxide into glucose (sugar) which it uses as food and releases oxygen for us to breathe. This process is called photosynthesis.

photo = light

synthesis = to make

Uses Light to Make Food and Oxygen

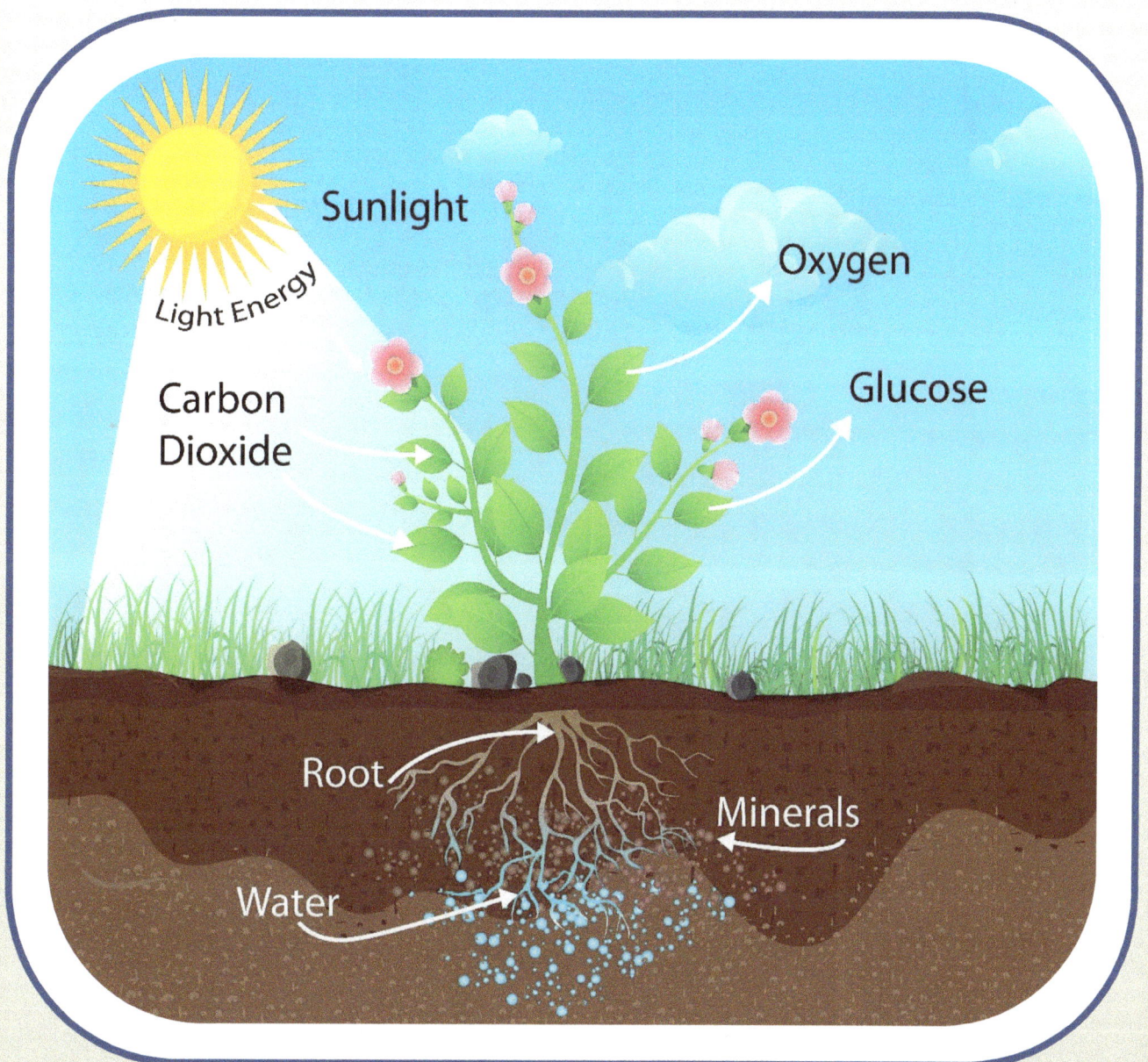

Sunlight

Light Energy

Carbon Dioxide

Oxygen

Glucose

Root

Minerals

Water

2

Tropical Rainforests

Tropical Rainforests

The **Climate** of a tropical rainforest is very hot. They are also **humid**. This means that there is a lot of moisture in the air. They are found in the tropic zone, near the equator. This is between the Tropic of Cancer and the Tropic of Capricorn (see Diagram 2 on p. 12). There are no seasons here, just hot and hotter.

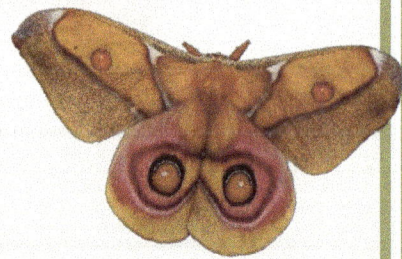

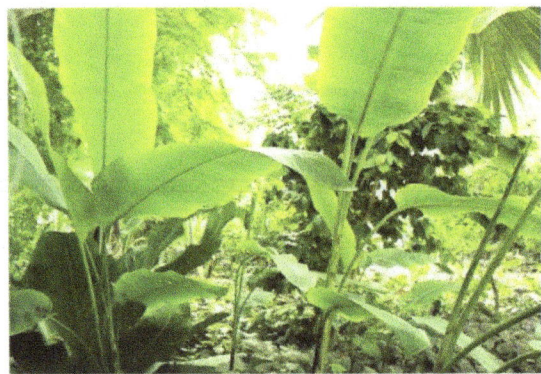

In tropical rainforests, you will find many varieties of broad leaf plants. Broad leaf means they are wide and flat. The **decomposition** rate here is fast because of the combination of heat and moisture which creates the perfect environment for **bacteria** and **fungi** to do their work breaking down dead material.

Epiphytes, such as orchids, ferns, and **bromeliads**, cover the branches of the trees. Liana vines hang like ropes, entangling themselves around the forest canopy. Large triangular buttress roots support the tall trees that grow in tropical rainforests. Most plants here stay green all year long. This is because they are not exposed to cold temperatures, like temperate rainforests. In temperate forests, trees become **dormant** or hibernate in the winter. This causes them to lose all their leaves.

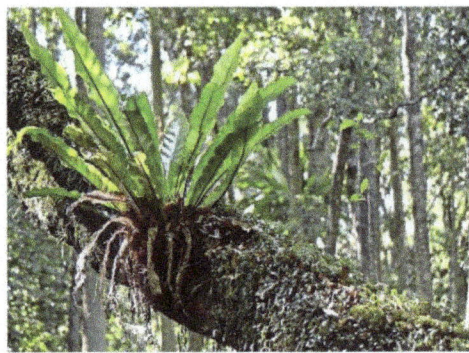

epiphyte
(e-pə-ˌfīt) a plant that derives its moisture and nutrients from the air and rain and grows usually on another plant

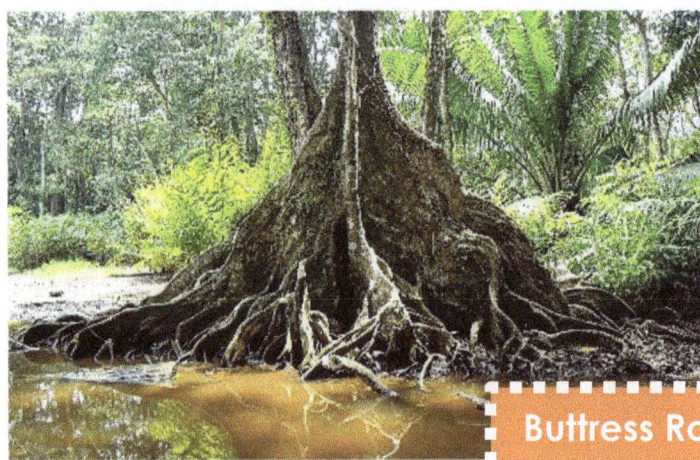

Buttress Roots

A rainforest **biome** is home to thousands of **species** of animals and plants. A biome is a naturally occurring community of plants and animals living in an area with a unique climate and **geology**. Each biome displays its own biodiversity, a variety of living things perfectly suited for life there. There is a desert biome, an aquatic biome, a rainforest biome, a grassland biome, and a tundra biome.

Tundra Biome

Desert Biome

There are many tropical rainforests all over the world filled with fascinating species and bizarre plants. We will be learning about 6 major tropical rainforests: Central American Rainforests, The Amazon Rainforest, The Congo Rainforest, The Madagascar Rainforest, SE Asian Rainforests, and Australian Rainforests. Even though our adventure is limited to pictures in a book, I hope someday you will see, touch, hear and experience these rainforests for yourself.

FUN FACT! **The Great Hornbill Bird** is only found in SE Asian Rainforests

A home in The Amazon Rainforest.

A baby anteater on its mommy's back.

FUN FACT! A Jungle is an area of land with dense, thick, and tangled vegetation. A rainforest can be described as a jungle but a jungle cannot always be called a rainforest.

The Galago (also called bushbaby) is only found in Africa and the Congo Rainforest. The head and body length is about 6-7 inches. It makes sounds that mimic the cry of a baby. It eats fruit and insects.

Actual Size

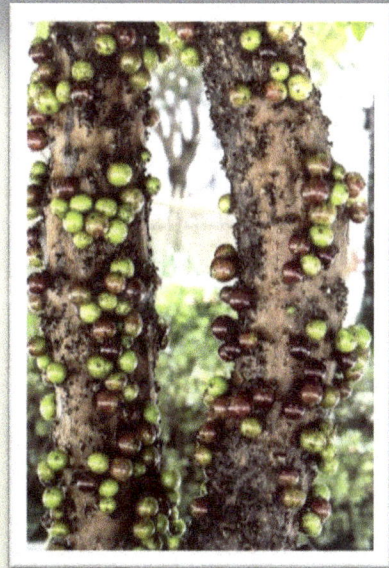

FUN FACT!

Jabuticaba is a small grape-size fruit found only in Brazil and the Amazon Rainforest. You pronounce it JA-BEW-TA-KABA (zhə- bü-ti-ˈkä-bə). It is also known as the Brazilian grapetree. This grape-like fruit grows right on the trunk and covers the branches in polka-dot clusters.

Australia

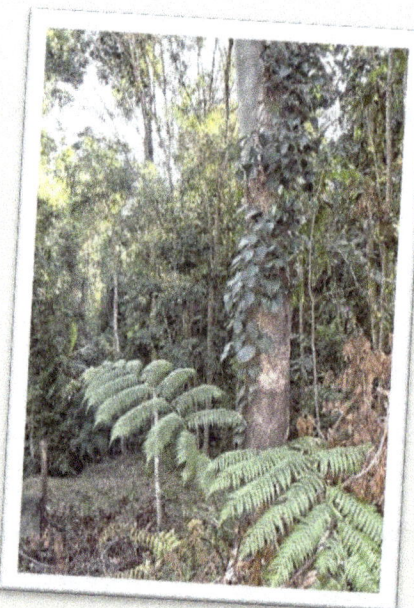

Australia is the name of the continent as well as the name of the only country found here. English is the primary language spoken. There are two tropical rainforests here and they are right next to each other. They are called the Daintree Rainforest and Kuranda Rainforest. They are found along the northeastern coast in Queensland. They lie along the Great Barrier Reef making it a popular place to travel because you can see the amazing tropical forests then go scuba diving to see the beautiful coral reefs and fish life in the ocean. Tasmania is a small island just south of the continent. This is where Australia's temperate rainforests are found.

PHOTO: © Borislav Marinic/123RF.COM

Kuranda is a town surrounded by a Rainforest. The beautiful Barron Falls is a popular place to visit here.

The Daintree Rainforest lies within the borders of a National Park. This is a picture of the Diantree River Ferry taking passengers across the water.

Map of Australia

TIMOR SEA
ARAFURA SEA
Torres Strait
CORAL SEA

INDIAN OCEAN

Darwin
Katherine
Kununurra
Derby
Halls Creek
Gulf of Carpentaria
Weipa
Coen
Cooktown
Cairns
Innisfail
Townsville
Ayr

Kuranda & Diantree Tropical Rainforests

New Caledonia (France)

Great Sandy Desert
NORTHERN TERRITORY
Mount Isa
Cloncurry
Normanton
Hughenden
Winton
Great Barrier Reef
Mackay
Rockhampton
Gladstone
Bundaberg

Port Hedland
Newman
Gibson Desert
WESTERN AUSTRALIA
Alice Springs
Simpson Desert
QUEENSLAND
Charleville
Roma
Sunshine Coast
Brisbane
Gold Coast
Ballina

Great Victoria Desert
SOUTH AUSTRALIA
Coober Pedy
Marree
Lake Eyre
Broken Hill
NEW SOUTH WALES
Dubbo
Bathurst
Newcastle
Coffs Harbour
Port Macquarie

Nullarbor Plain
Ceduna
Lake Torrens
Lake Gairdner
Whyalla
Port Augusta
Port Pirie
Sydney
Wollongong

Perth
Mandurah
Bunbury
Great Australian Bight
Adelaide
VICTORIA
Bendigo
Ballarat
Melbourne
Geelong
CANBERRA
Goulburn
Cape Howe

Albany
Temperate Rainforest
Bass Strait
PACIFIC OCEAN

INDIAN OCEAN
Tasmania
Launceston
Hobart
TASMANIA
TASMAN SEA

SLOW DOWN

Dingo

FUN FACT! **Cassowary birds** are Australia's largest land animal!

23

CASSOWARY

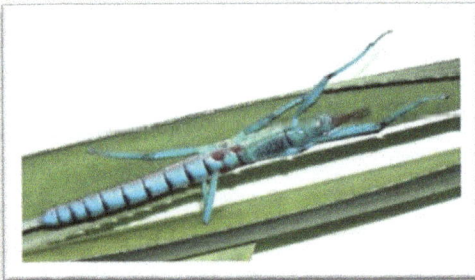

Peppermint stick bug

If this insect is disturbed it will spray a milky substance that smells like peppermint. These blue-green bugs are almost 4 ½ inches long. They live and feed on the foliage of the screwpine tree (Pandanus tecturius).

Amethystine Python

(a-mə-ˈthis-tən) This snake can grow up to 28 feet long making it the sixth largest snake in the world. Its name comes from the color of its skin which has a purple hue, also called amethyst.

PYTHON PHOTO: By One dead president - Own work, CC BY-SA 3.0, https://commons.wikimedia.org/w/index.php?curid=4078818

prehensile- (prē-ˈhen(t)-səl) capable of grabbing or holding something by wrapping around it

The Striped Possum

This animal looks like a mix between a squirrel and a skunk. A **prehensile** tail is used to hang from branches to reach food. It has also been featured on Australian stamps.

AUSTRALIA
50c
striped possum

© Borislav Marinic/123rf.com

Red-legged Pademelon

These small rainforest kangaroos stand at a height of only 2 ½ feet tall.

Echidna (i-ˈkid-nə) This spiny-coated animal has no teeth. It uses its long sticky tongue to eat ants and insects. Echidnas and platypuses are the only egg-laying mammals in the world.

Cruiser Butterfly

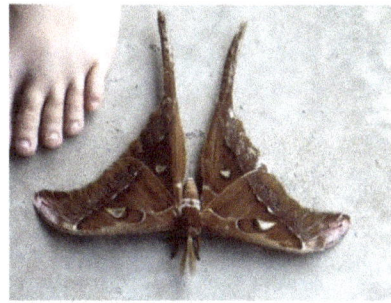

Hercules Moth

These moths have a wingspan that measures just over 9 inches. They are often spotted hanging around the bleeding heart tree.

Galah

Rainbow Lorikeet

Australian King Parrot

25

The plants in Australia's rainforests have some very unusual names!

Quandong Tree

Elephant Ears

The Stinging Tree

This is the most poisonous plant in Australia. It is also known as the gimpy stinger. Little poisonous hairs cover the leaves and stems. If you were to brush up against it, these hairs would penetrate your skin causing an extremely painful burning sensation. In severe stings this pain can last up to 6 months.

PHOTO By Cgoodwin - Own work. CC BY 3.0.

Hairy Gardenia

These bright white flowers standout in the dark understory of Australia's rainforests. This small shrub is covered with hairy leaves.

Bleeding Heart

Striped Cucumber

Strangler Fig Tree

The fruit of this tree is eaten by birds. The seeds are scattered in their droppings from high in the canopy. The seeds will then sprout and send roots down into the ground which gives it more nutrients and allows it to grow more rapidly. This plant will wrap itself around the host tree and begin to strangle it slowly, eventually leading to its death.

Lemon Myrtle

Lemon Myrtle is **native** to Australia's rainforests. The leaves have a strong lemon scent which is used as a spice to season food such as fish, chicken, or beef. It is also used to make tea, lotion, shampoo, conditioner, and essential oil.

Ginger

Ginger is a spice. The root is the part of this plant that you eat. If you like ginger ale soda, pumpkin pie, or ginger bread cookies then you have most likely eaten ginger. It also grows abundantly in Asia.

Riberry

Its pronounced rye-berry. This shrub is native to Australia. The berries taste like a mix of cranberries and cloves. It is commonly used to make jam.

Eucalyptus

This tree provides timber for building.

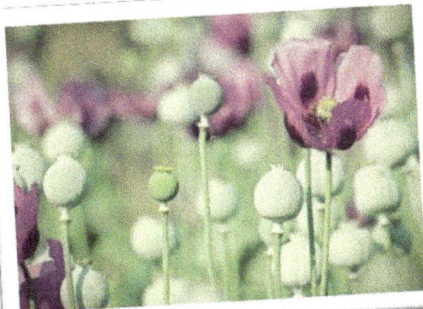

Poppy Flowers

Australia is one of the few countries that grows poppy flowers. Poppies have a drug in them that is used as a pain killer by doctors and hospitals around the world.

South America

The Amazon is the largest rainforest in the world. It is located on the continent of South America. It extends into three countries: Brazil, Columbia, and Peru. The Amazon River bends and twists through this vast tropical **jungle**. The Yanomamo (ya-no-mom-o) people live in this rainforest and use its natural resources to survive and live off the land.

Macaws

Amazon River

CARIBBEAN SEA

ATLANTIC

OCEAN

Cristóbal Colón Peak
Barranquilla
Cartagena
Maracaibo
CARACAS
Cabimas
Barquisimeto
Mérida
Bolívar Peak
Maturín
CARACAS
Ciudad Bolívar
Ciudad Guayana
VENEZUELA
GEORGETOWN
Medellín
PARAMARIBO
Manizales
BOGOTA
Cali
COLOMBIA
Neiva
GUYANA
SURINAME
FRENCH GUIANA
Cayenne
Popayán
Pasto
Inírida
Boa Vista
Esmeraldas
ECUADOR
QUITO
Portoviejo
Chimborazo
Guayaquil
Cuenca
Mitú
Rio Negro
Amazon
Belém
São Luís
Sullana
Piura
Iquitos
Leticia
Amazon
Manaus
Santarém
Teresina
Fortaleza
Chiclayo
Trujillo
Rio Branco
Rio Branco
B R A Z I L
Mossoró
Natal
João Pessoa
Huascarán
Porto Velho
Palmas
Recife
Maceió
LIMA
Cuzco
Juazeiro
Aracaju
PERU
BOLIVIA
Cuiabá
Salvador
Lake Titicaca
Itempu
Anápolis
BRASÍLIA
Itabuna
LA PAZ
Corabelamba
Goiânia
Vitória da Conquista
Arequipa
Arica
SUCRE
Santa Cruz
Campinas
Governador Valadares
Iquique
Potosí
Tarija
Corumbá
Campo Grande
Belo Horizonte
Vitória
Calama
PARAGUAY
Pedro Juan Caballero
São Carlos
Antofagasta
Salta
Concepción
Campinas
Rio de Janeiro
Salta
ASUNCIÓN
São Paulo
Santos
San Miguel de Tucumán
Posadas
Curitiba
Resistencia
Corrientes
Florianópolis
La Rioja
Santiago del Estero
Aconcagua
Córdoba
Santa Fe
Rivera
URUGUAY
Rio Grande
Valparaíso
Mendoza
Rosario
Paraná
Mercedes
Porto Alegre
SANTIAGO
Rancagua
Rio Cuarto
BUENOS AIRES
MONTEVIDEO
Pampas
La Plata
Concepción
ARGENTINA
CHILE
Temuco
Neuquén
Mar del Plata
Bahía Blanca
Puerto Montt
Viedma
San Carlos de Bariloche
Rawson
Patagonia
Comodoro Rivadavia
SOUTH
PACIFIC
OCEAN
SOUTH
ATLANTIC
OCEAN
Falkland Islands (UK)
Stanley
Puerto Natales
Rio Gallegos
Punta Arenas
Ushuaia
Cape Horn

29

Jaguar

These **carnivores** are the largest cats in South America. They can live up to 15 years. Their black spots look like roses.

Carnivore- (kär-nə-ˌvor) an animal that feeds on meat

Sloth

There are two kinds of sloths: two-toed and three-toed. Because of their long claws, moving on the ground is difficult so they spend most of their time hanging in the trees. These slow-moving mammals can sleep for up to 20 hours a day.

Curly Crested Aracari

Giant Anteater

This long-nosed animal has no teeth. It uses its tongue to lick up thousands of ants every day. The giant anteater can grow up to 7 feet long from the tip of its nose to the end of its tail.

Poison Dart Frog

Golden Lion Tamarin

The golden lion tamarin is a tiny monkey. It can grow 6 to 10 inches in length and less than 2 pounds in weight. They sleep in holes in trees at night. They are **omnivores** and consume a diet of fruit, bugs, amphibians, reptiles, birds, flowers, **nectar**, and eggs.

Kinkajou This animal can rotate its ankles backwards which allows it to move more freely throughout the trees.

Capybara

This rainforest is home to the largest rodent in the world called capybara. It can grow up to 2 feet tall and weigh up to 100 lbs.

Glass Winged Butterfly

Just like its name implies, their wings are see-through.

Uakari
(wä-ˈkä-rē)

Harpy Eagle

Armadillo

Toucan

Sapodilla Tree

This tropical evergreen tree is found in Mexico and Central America. It has brown round fruit that tastes like a mix of pear and brown sugar. A milky latex called chicle is extracted from the trunk by tapping. This is where you cut a slit in the tree and catch the sap that flows out. The sap is used to make gum.

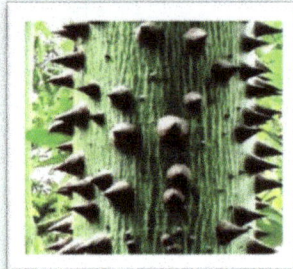

Kapok Tree

This is a giant tree that can reach heights of 200 feet. It is covered in thorns for the first year of growth. The wood has been used to make dugout canoes. The pink flowers smell bad and attract bats. Each fruit or seed pod contains about 200 seeds surrounded by fiber. The seeds have an oil in them that can be used to make soap. The fiber is used for stuffing cushions and pillows.

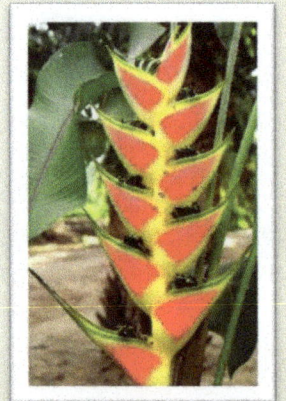

Orchid

Giant Lily Pad

Passiflora

Heliconia

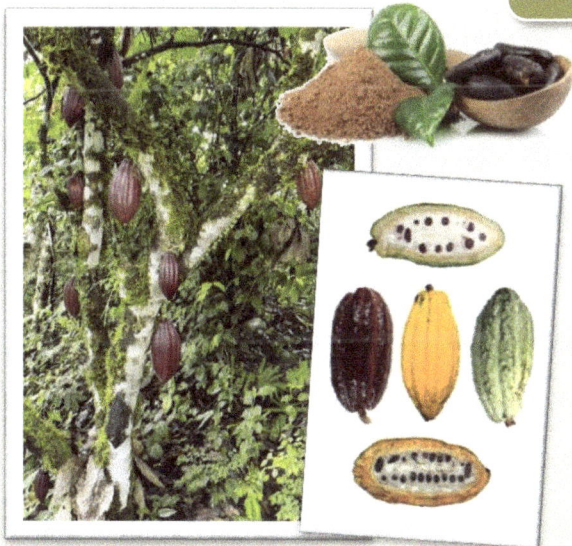

Chocolate

Chocolate comes from the seed of a fruit. They grow on the cacao tree which is pronounce ka-kow not ko-ko like we call the powdery stuff in hot chocolate. The football-shaped fruit pods grow to about 6 inches in length. The edible white flesh around the seeds has a sweet lemony flavor. The seeds don't get a chocolate flavor until they are fermented, dried, and roasted. Then they are ground up and made into chocolate by adding sugar and milk.

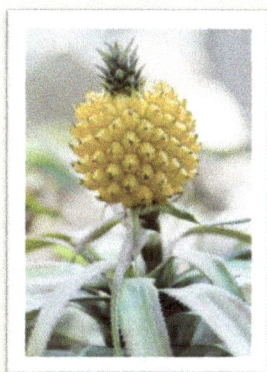

Coffee

Coffee comes from the coffee tree which can grow up to 15 feet tall. The branches are covered with bright red berries. There are usually 2 beans inside each berry. Coffee is made from these beans through a roasting process.

PINEAPPLE

Poinsettia

Poinsettia grows abundantly in the Amazon and is used for decoration at Christmas time in America.

Acai Palm

(ä-ˌsä-ˈē)
This tree provides a healthy fruit called acai berry. This palm grows in the amazon rainforest.

Brazil Nuts

Brazil nuts grow inside the fruit of the Bertholletia excelsa tree.

Kapok Fiber

The kapok tree produces a cotton-like fiber and is often used for stuffing cushions.

Africa: Congo

The Congo Rainforest lies in The Congo Basin. This large rainforest extends into six of Africa's countries: The Democratic Republic of Congo, The Republic of Congo, Equatorial Guinea, Gabon, The Central African Republic, and Cameroon. Those are some long and strange country names so don't let them confuse you. The Congo River runs through these jungles and is one of the largest in the world.

The Pygmy tribal people live in these forests. They spend their time farming, hunting, and surviving off the land. They are an interesting people group because they are the smallest people in the world. Their average height is about 4 ½ feet tall.

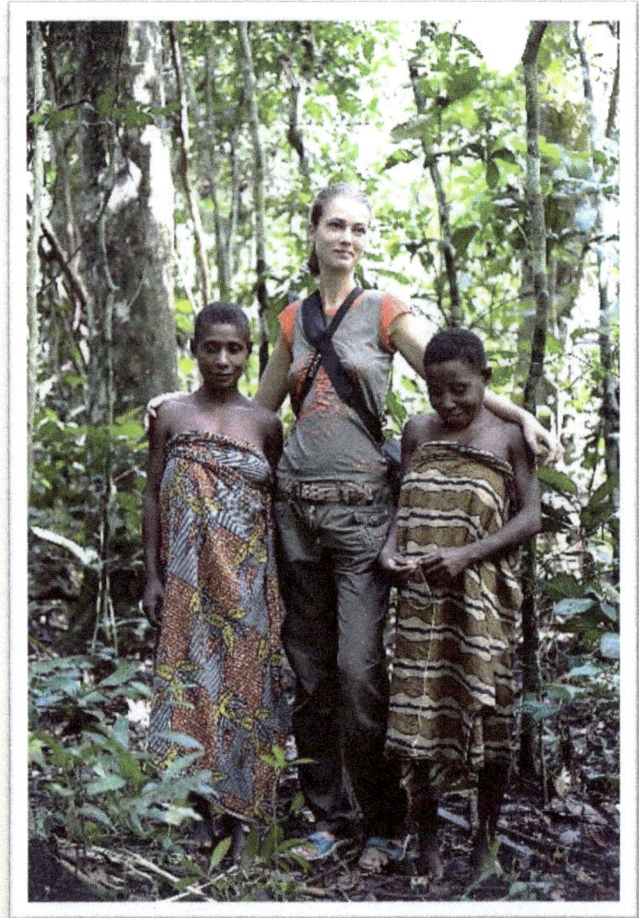

Photo provided by Sergei Uriadnikov Copyright: <a href='https://www.123rf.com/profile_surz'>surz / 123RF Stock Photo</a>

Mandrill

35

Colobus Monkey

These monkeys don't have thumbs like other monkeys do. Their babies are born with soft white fur. They can jump up to 19 feet through the air from tree to tree. Even a mommy monkey can jump this far with a baby on her back. They eat flowers, leaves, twigs, and fruit.

African Forest Elephant

Forest elephants inhabit the dense wooded areas of the Congo rainforest. They are much smaller than other African elephants. Their tusks are straighter and point downward unlike savanna species. They are herbivores and eat leaves, fruit, and bark.

Okapi

This animal looks like a mix between a giraffe and zebra. They can grow up to 8 feet tall. Its velvety fur is a reddish-brown color with horizontal stripes on its legs. This strange animal has a prehensile tongue that is 18 inches long. They use it to hold and strip leaves from stems. Their diet includes a variety of fruits, ferns, leaves, and twigs.

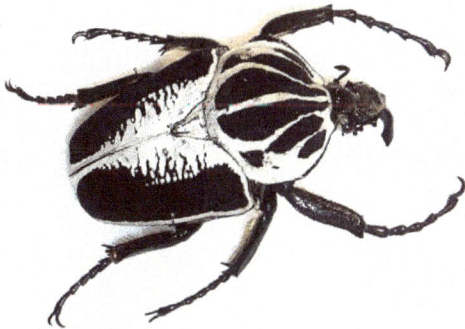

Goliath Beetle

The Congo rainforest is home to the largest beetle in the world. It can grow up to 4 inches in length.

Bonobo

The bonobo, also called the pygmy chimpanzee, spends most of their time on the ground. With long legs, and narrow shoulders these animals can walk upright standing on two legs. They are between 29 and 35 inches tall and weigh between 55 and 100 pounds.

Pygmy Hippopotamus

These hippos live in the Congo River.

Gorilla

Gorillas move around by knuckle-walking. They make nests in trees or on the ground to sleep in. They eat fruits, leaves, and shoots.

African Grey Parrot

Giant African Millipede

Gasso Nut

Climbing high on the trees, this plant produces stems that can grow up to 90 feet in length. The nutritious seeds are harvested by local people and are sold in markets. This plant is also used as medicine.

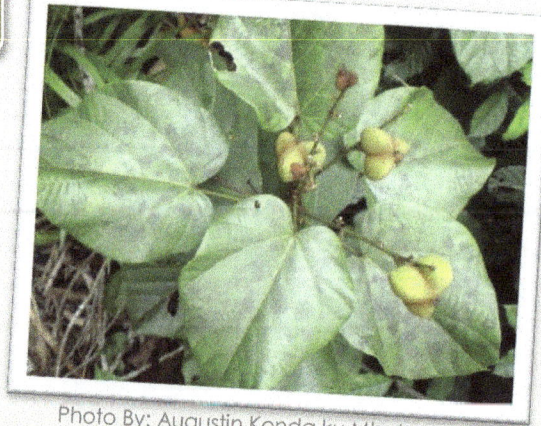

Photo By: Augustin Konda ku Mbuta

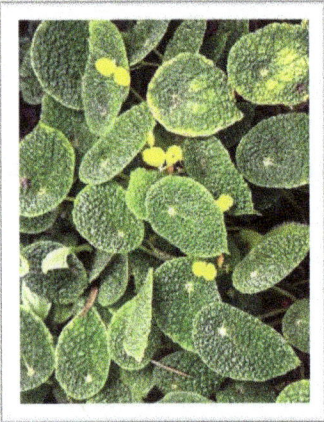

Begonia Microsperma

This plant is native to the Congo Basin area. The bright yellow flowers stand out against the deep green leaves.

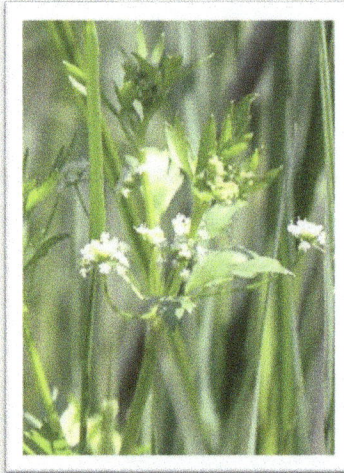

Wild Celery

This plant grows in the Congo rainforest. Wild celery is part of the gorilla's diet.

African Impatiens

Commelina

FUN FACT!

Bromeliads have a cone-like center which provides water for canopy-dwelling animals to drink. Frogs also use them for water habitats high up in the trees.

Moabi Tree

This tall tree stands high above the canopy layer. It is a popular hardwood tree used for building.

African Palm Oil

The berry from this plant is harvested for its oil which is called palm oil. It's used in many products such as chocolate, candles, lotion, and soap.

Gold

Cassava

This is an important food in the tropics. It is similar to a sweet potato. The root or tuber is the part you eat. It is also called yuca and when dried into a powdery or pearly extract it's called tapioca.

African Teak

This **deciduous** tree grows abundantly in the Congo rainforest. It provides quality lumber for building boats and homes.

deciduous

(di-ˈsi-jə-wəs) having leaves that fall off every year

Africa: Madagascar

Madagascar is an island and country off the east coast of Africa. It is home to some spectacular plants and animals. Lemurs are a species of animals that are **endemic** to this rainforest. Endemic means they are only found in this area. There are no large mammals on this island, only small ones such as fossas, hedgehogs, and shrew tenrecs.

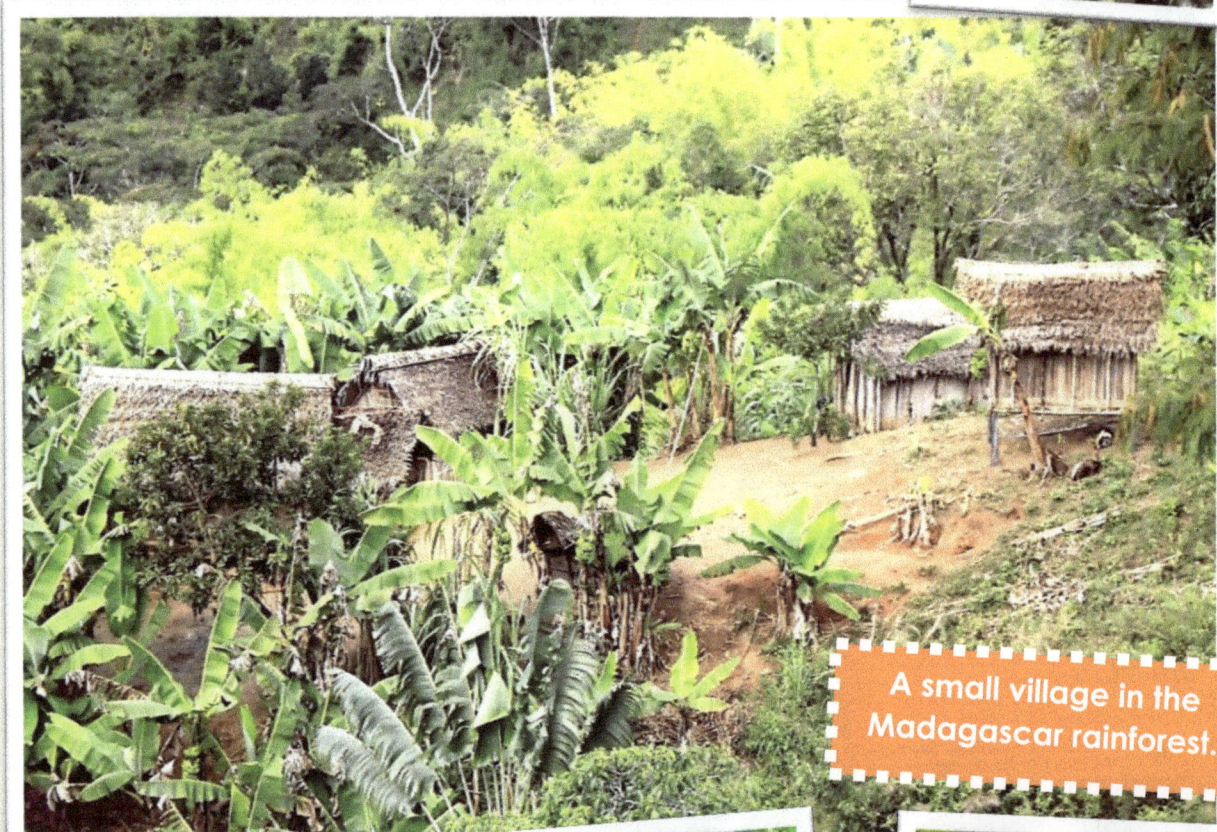

A small village in the Madagascar rainforest.

Striking buttress roots found inside the Masoala National Park.

Grande Comore

Mohéli   Anjouan

COMORES

Mayotte (Fr.)

Glorioso I. (Fr.)

Cap d´Ambre

Antsiranana

Nosy Mitsio

Ambilobe

DIANA

Vohemar

Nosy Be

Ambanja

SAVA

Bealanana

Sambava

Andapa

Antsohihy

Antalaha

SOFIA

Maroant-setra

Mahajanga

Mandritsara

ANALAN-JIROFO

MOZAMBIQUE

Cap Saint André

Soala

Lake Kinkony

Marovoay

BOENY

ALAOTRA-

Soanierana Ivongo

Nosy Sainte Marie

Maevatanana

BETSIBOKA

Ampara-faravola

Fenoarivo Atsinanana

Mananara Avaratra

Juan de Nova I. (Fr.)

MELAKY

Lake Alaotra

Ambatondrazaka

MANGORO

Toamasina

Maintirano

ANALA-MANGA

ATSINA-

Nosy Barren

Antsalova

BONGOLAVA

Manjakan-driana

Tsiroano-mandidy

Miarinarivo

ANTANANARIVO

MADAGASCAR

ITASY

NANA

Belo Tsiribihina

Faratsiho

Tsiribihina

VAKINANKARATRA

Antani-fotsy

Antsirabe

Mahanoro

Mania

Fandriana

Morondava

AMORON'I MANIA

Ambositra

Nosy Varika

MENABE

VATOVAVY-

Manja

INDIAN

MATSIATRA AMBONY

FITO-

Mananjary

Morombe

Fianarantsoa

Lake Ihotry

Mangoky

VINANY

OCEAN

Manakara

Europa Island (Fr.)

Ihosy

IHOROMBE

ATSIMO-

Farafangana

ATSIMO-ANDREFANA

ATSINA-

Toliara

NANA

Betioky

ANOSY

Lake Tsimanampetsotsa

Beraketa

ANDROY

Amboasary

Tolanaro

Ambovombe-Androy

Cap Sainte Marie

Africa

FUN FACT!   The official Languages of Madagascar are Malagasy and French.

41

Civet

This small cat-like animal can weigh up to 5 pounds. It is **nocturnal** and eats insects, rodents, reptiles, frogs, and even eggs it steals from birds' nests. It has a gland that produces a fluid which is used as a musk. A musk is a perfume that comes from an animal. The odor from this fluid is strong and putrid, however, when diluted it has a sweet and pleasant smell.

nocturnal- (näk-ˈtər-nᵊl) active at night

Chameleon

There are many varieties of chameleons found in Madagascar, from the miniscule 1-inch species to the colorful panther chameleon.

Fossa

The fossa can weigh up to 20 pounds. It has flexible ankles which allow it to climb up and down trees head first.

Sunset Moth

Tomato Frog   Tenrec

Lemur

There are many different species of lemurs: ring-tailed, gray mouse, sifaka, and the aye-aye. The sifaka species is unique because it can move quickly on the ground by doing a two-legged sideways hop. They are mostly nocturnal but can be active during the day. Fossas are their natural predator.

Baobab Tree

This bizarre tree can grow to a height of 98 feet tall with a diameter between 23 and 36 feet. It looks like it is upside down with root-looking branches. The trunk of this tree swells up with water.

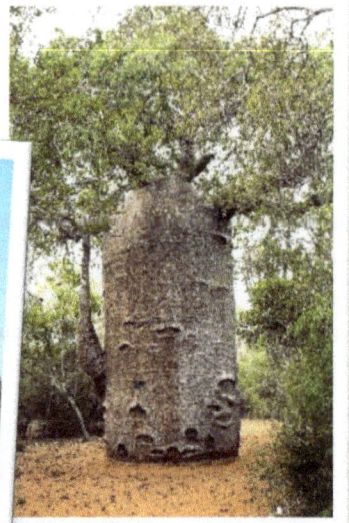

Traveler's Tree

This fan-like tree is found all over Madagascar. It has enormous paddle-shaped leaves and large white flowers.

Alluaudia

The stems of this plant are covered in spines. Lemurs love to eat them.

Orchids

Bat Flower

Vanilla Beans

Vanilla is the most popular ice-cream flavor in America. It comes from a plant called Vanilla Bean Orchid which is an epiphyte. It's the only orchid plant that produces a fruit. This rainforest is known for growing the desirable Madagascar Vanilla Bean.

Raffia Palm

This tree is very useful. Raffia fiber comes from the veins of the leaves and is used in crafts or made into twine, baskets, mats, shoes, and hats. The large leaf stalks are used to build houses and furniture. A wax is extracted from the leaves and is used on floors and as a shoe polish.

Chocolate

The cocao trees here have been known to produce the best quality of beans for chocolate products throughout the world.

Mango

Periwinkle

Sweet Potato

Asia

There are many different rainforests scattered throughout the south-eastern region of Asia, we call them the Southeast Asia Rainforests. They are located in the coastal countries of Myanmar, Thailand, Cambodia, and Malaysia. They are also found on the islands of Indonesia and the Philippines. These forests are home to many unique-looking animals and fascinating plants. Animals such as the orangutan, silvery gibbon, and the slender loris are only found in these jungles.

Thailand Rainforest

SOUTHEAST ASIA RAINFORESTS

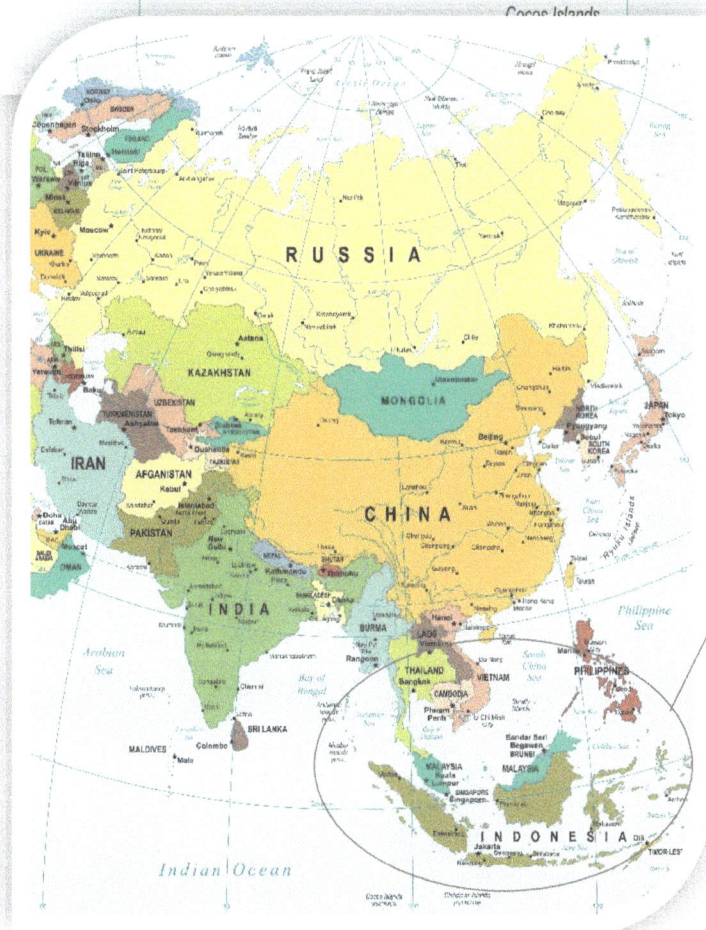

Hornbill

Proboscis Monkey

The **proboscis** monkey has an unusually large nose. Males have a bigger nose than females. To communicate, they will honk, roar, and snarl at each other. They are good jumpers and swimmers. Their diet includes a variety of fruit and leaves.

proboscis

(prə-ˈbäs-əs, -kəs) the long, thin nose of some animals

Malayan Tapir

Orangutan

Orangutans have reddish-brown hair, bulky bodies, bowed legs, and strong arms. Dominant males have distinctive cheek pads that give them a strange circular face shape. They spend most of their time in trees and making elaborate sleeping nests. Fruit makes up most of an orangutan's diet.

Tarsier (tär-sē-ər)

Big eyes and a small body give this creature a very interesting look. They have long fingers and soft, velvety fur. Their brain is the same size as one of their eyeballs. These nocturnal animals feast on insects, snakes, and lizards.

Goliath Birdwing

This is the second largest butterfly in the world. It has a wingspan of up to 11 inches. The Queen Alexandra's Birdwing is the largest species with a 12-inch wingspan.

Sunbear

Flying Fox

Bengal Tiger

49

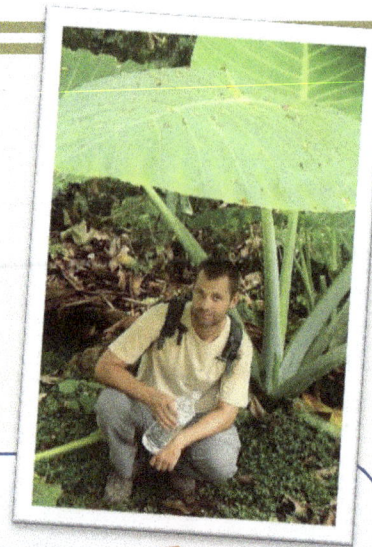

Borneo Giant

The massive leaves of this plant can grow up to 5 feet wide. If eaten, it will cause numbness and pain in the tongue and throat. This is followed by swelling which will make it difficult to breathe.

Sugar Cane

Rafflesia

Hibiscus

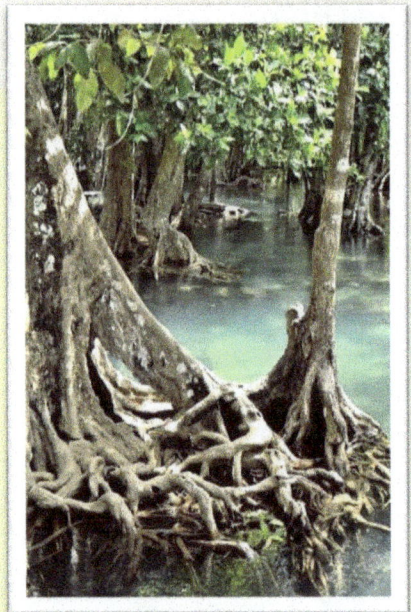

Titan Arum

The fragrance of this 10-foot-tall flower resembles rotting meat. This is why it has the nickname of "corpse flower". The foul smell attracts insects which pollinate it.

Pitcher Plant

At the tip of the sword-shaped-leaves emerges a tube-like trap. The trap has a syrupy fluid in it that is used to drown the **prey** of this carnivorous plant.

Mangrove

Mangroves are a small tree that grow in salty water. They have long, complex roots to help them survive the harsh coastal conditions.

Durian

This tree species grows a fruit with a thorny husk. The fruit can grow up to 12 inches long and 6 inches in diameter. It has a very strong odor. Some people say its sweet and pleasant, others find the aroma overpowering. The flesh has an almond-custard flavor. The tree can grow to a height of 164 feet.

Ylang-ylang

This plant is used to make medicine, essential oils, and perfume.

Bananas

Many varieties of bananas grow in Asia.

Sugar Cane

Sugar cane is native to Asia. However, it is now widely grown in many tropical countries including the Amazon Rainforest.

CINNAMON

PEANUTS

PEPPER

CLOVES

North America

Central America is located in the southernmost portion of the North American continent. There are many rainforests found here. These rainforests are located in Southern Mexico, Panama, Costa Rica, Honduras, Belize, and on some of the Caribbean islands. The cities which are surrounded by crystal clear blue water, warm weather, and tropical rainforests are popular destinations for vacationers. The narrow stretch of land that connects North and South America is home to some amazing plants, animals, and spectacular sights. From the San Luis waterfalls to views of beautiful volcanos, these rainforests have a scenery like nowhere else in the world. Screeches of the mantled howler monkey can be heard for miles. The famous poison dart frog is also found in these jungles.

Motmot

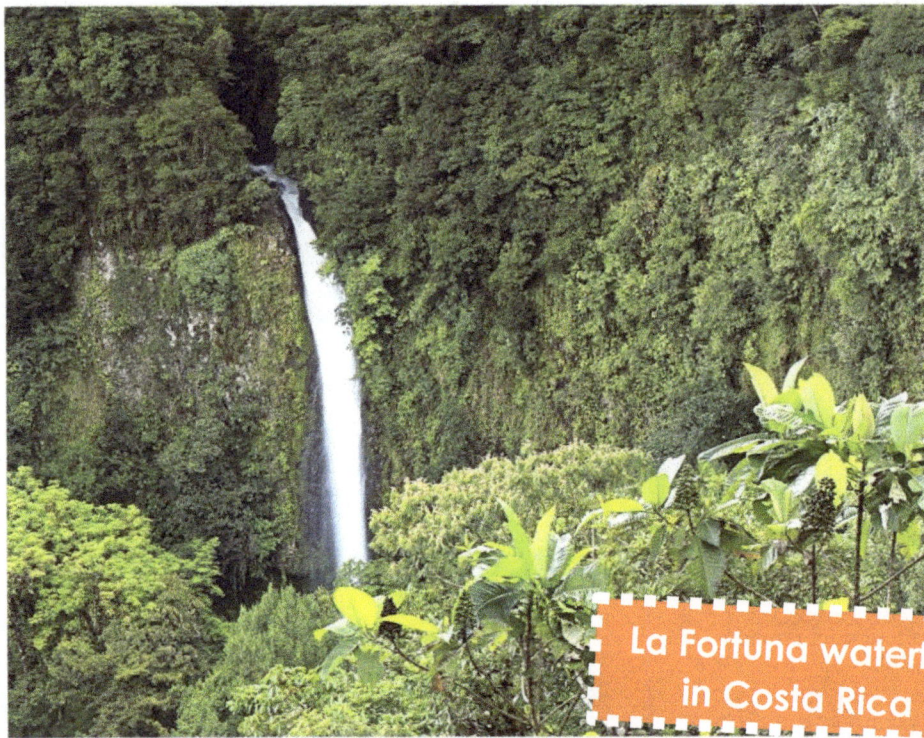

La Fortuna waterfall in Costa Rica

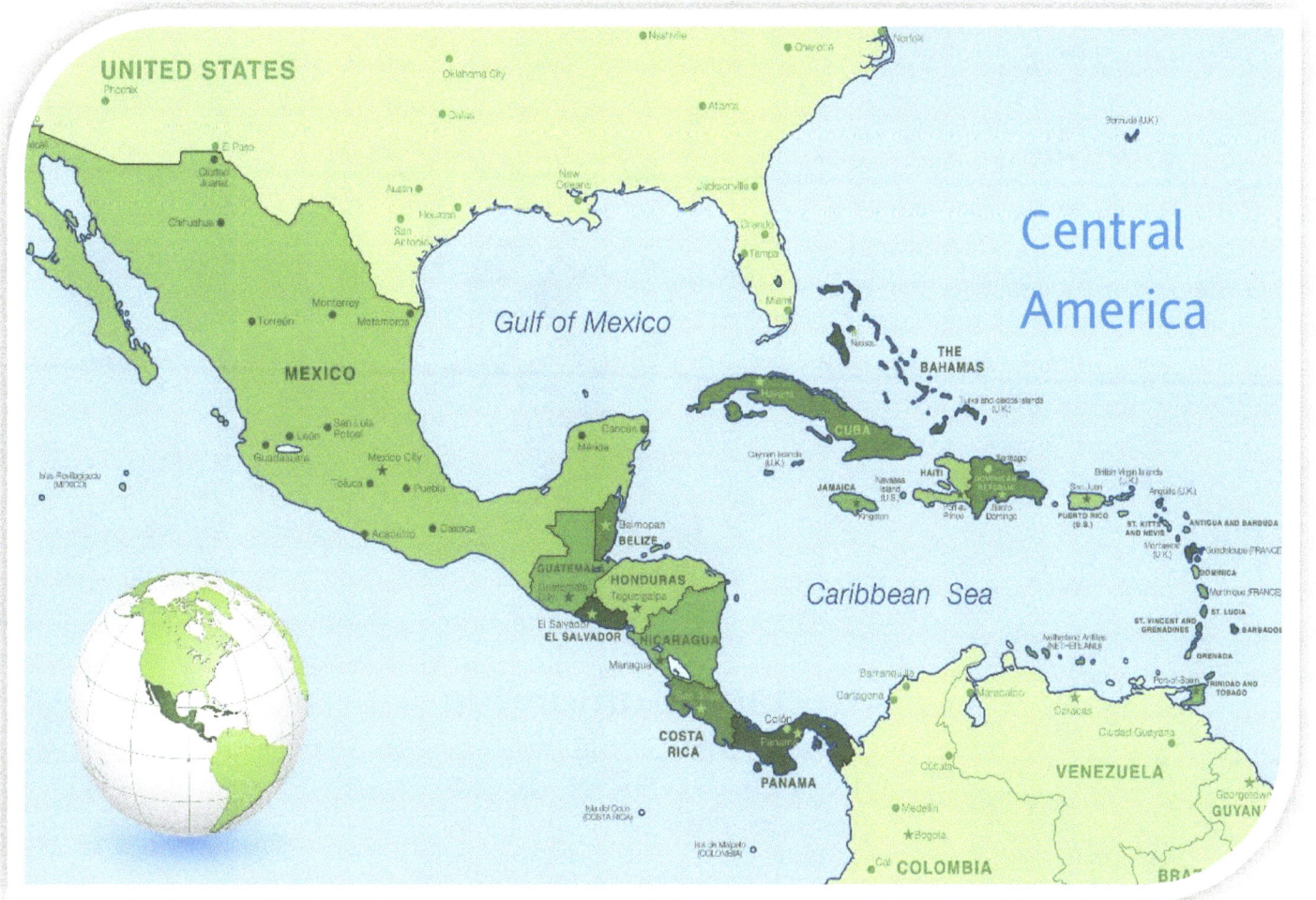

Central America

Gulf of Mexico

Caribbean Sea

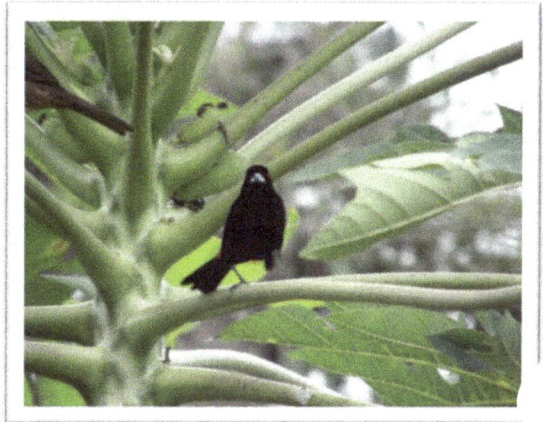

Capuchin (ka-pyə-shən)

These clever monkeys have been observed using plants as medicine by rubbing leaves all over their bodies to keep away bugs. They also use tools tor weapons and to get food. They are a medium sized monkey with a prehensile tail.

Hummingbirds

Quetzal

Northern Tamandua

These muscular anteaters have an unusually large claw on the middle toe of their forefeet. They use these to rip open stumps and trees to find insects. They lap them up with their long sticky tongues. These anteaters grow up to 51 inches in length and 12 pounds in weight.

Margay This small leopard-like cat is native to Central and South America. It is very similar to the ocelot. It has a body length that measures between 19-31 inches. The margay spends most of its life in trees.

Poison Dart Frog

This frog gets its name from the **indigenous** people who used its toxic secretions to poison the tips of their blow darts. These frogs come in a variety of colors.

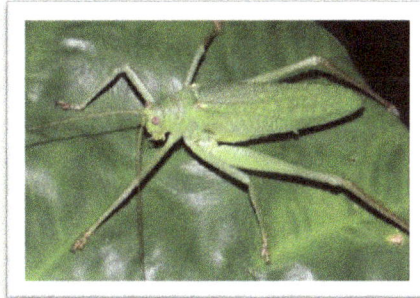

Katydid

This cricket gets its name from the song it sings. It sounds like "katydid-katydidnt". Its leaf-like body can grow up to 5 inches in length.

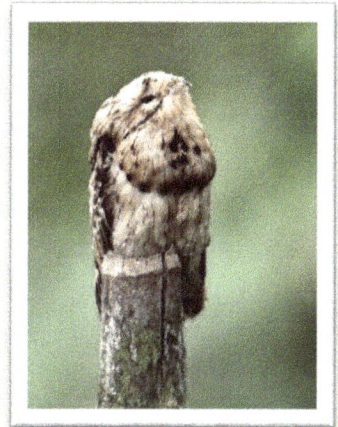

Potoo

This nocturnal bird will **camouflage** itself during the day by perching on the tops of tree stumps.

BASILISK

HOWLER MONKEY

SLOTH

Spider MONKEY

Cannonball Tree

From beautiful blooms to the 8-inch cannonball looking fruit, this impressive tree will not disappoint your eyes. Some trees produce up to 1,000 flowers at one time, covering the entire tree. Many animals feed on the fruit pulp and seeds. Although edible, people usually do not eat this fruit because it has an unpleasant smell. It is also used to make medicine.

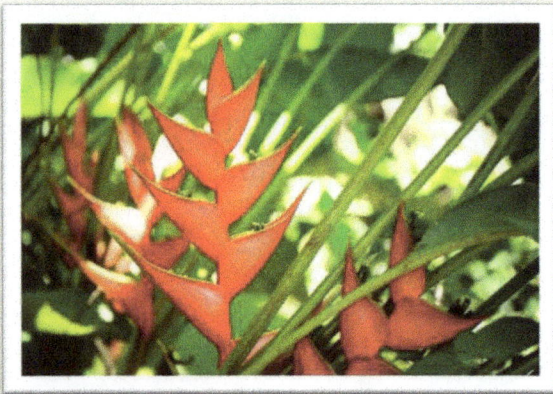

Heliconia

The heliconia has bright alternating flower blossoms. It produces a sweet nectar that attracts pollinators such as the hummingbird.

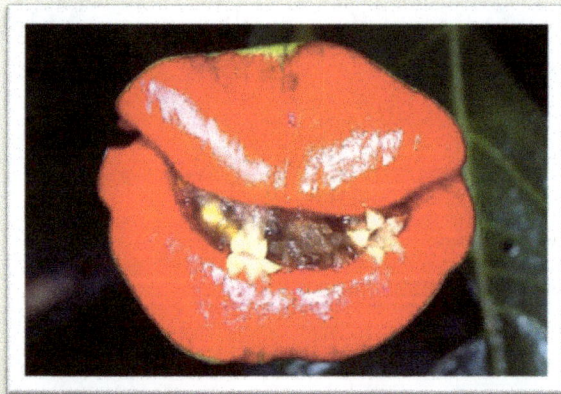

Photo by Morley Read/123rf.com

Psychotria

This plant has flowers that look like red lips. It makes its home in the shadows of the understory.

PALM TREE COCONUTS

MONKEY LADDER LIANA

Panama Rubber Tree

Rubber is made from the milky **latex** extracted from this important tree. Without it we would not have tires, life-saving medical equipment, erasers, balloons, paint, inflatable boats...and many more things. People collect the latex by cutting a slit in the tree, then catch the liquid in a bowl.

Guava

This beautiful pink-fleshed fruit can grow up to 4 inches in diameter. It has a light, lemony fragrance. You can eat it fresh off the tree like an apple. Most often though, it is made into fruit punch, jams, and desserts.

Cashews

This tropical evergreen tree produces an apple and a nut. The tree can grow up to 46 feet tall. The 4-inch cashew apple is edible and is often made into a juice which tastes like mango with a hint of spicy citrus flavor. One cashew nut grows from one cashew apple. The shell around the nut has a toxin similar to poison ivy. Properly roasting it destroys this toxin.

Mahogany

Mahogany is a tree with straight-grained wood that is used to make furniture. It is prized for its beauty, durability, and color.

Chapter

3

Temperate Rainforests

Temperate Rainforests

All temperate rainforests are mild in temperature - not too hot and not too cold. They are located above and below the tropic zone. You will find them on the west coast of North America, on the southern tip of South America, by the Black Sea in Europe, and on the east coast of China. There is also a temperate rainforest in southern Australia on the island of Tasmania. These rainforests are considerably different from tropical rainforests. They are home to their own unique variety of animals and plants.

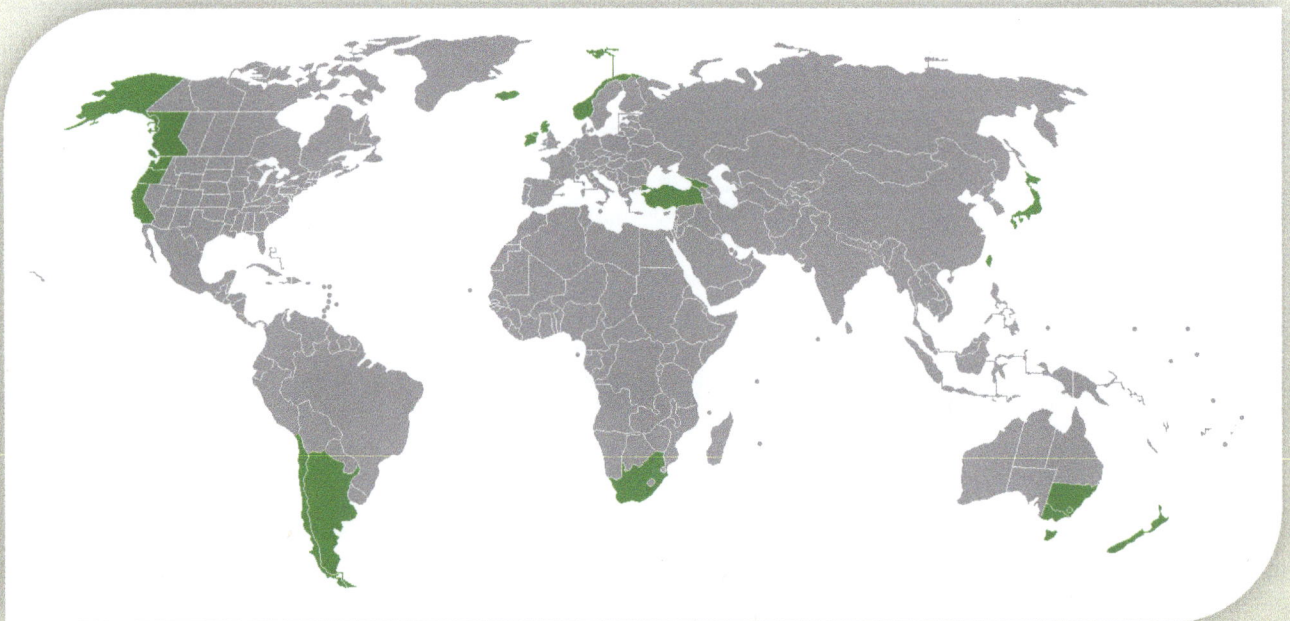

Animals

Unlike tropical rainforests, most animals here live on the forest floor. They have a variety of ways they can survive the temperature changes. Some **hibernate** in winter when food is scarce. Others store up food for the winter. Still others will **migrate** to warmer areas. Many bugs lay eggs that can withstand the cold temperatures. There are a small number of animals that remain active and are able to find food to survive.

migrate (mī-ˌgrāt) to pass usually periodically from one region or climate to another for feeding or breeding

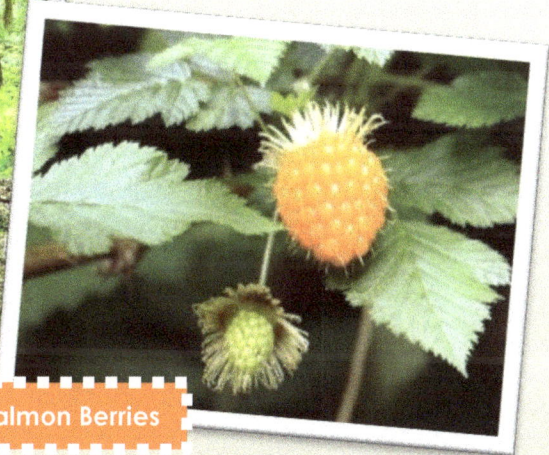

Salmon Berries

Trees and Plants

Evergreen trees stand high above the forest floor. Their branches are covered in needle-like leaves which stay green all year long. They produce a sap inside that keeps their roots from freezing. The sap also provides the tree with energy throughout the cold season. Many trees and plants become dormant and shed their leaves every year in autumn. This preserves energy so the plants can survive the winter. These are called deciduous.

Dead plant material and other decaying matter decompose very slowly here. This creates a forest floor that is covered with fallen trees and a thick layer of leaves. Epiphytes that grow here are mostly ferns and mosses. Not only do they grow on the trees but you will also find them living on all the decaying logs and **debris** (də-ˈbrē) on the ground creating a beautiful carpet of plants.

The Hoh Rainforest

Ginormous trees, strange animals, and drooping moss give this forest a jungle-like feel. The temperate rainforests of Washington are home to the slimy banana slug, the herbivorous stinkbug, and the striking roosevelt elk. Many mysteries of the Hoh rainforest lie beneath the branches of tall evergreen trees. From unique animal species to a **diverse** plant life, you will not be disappointed. Let's go on an adventure and take a closer look to find out what The Hoh Rainforest is really like.

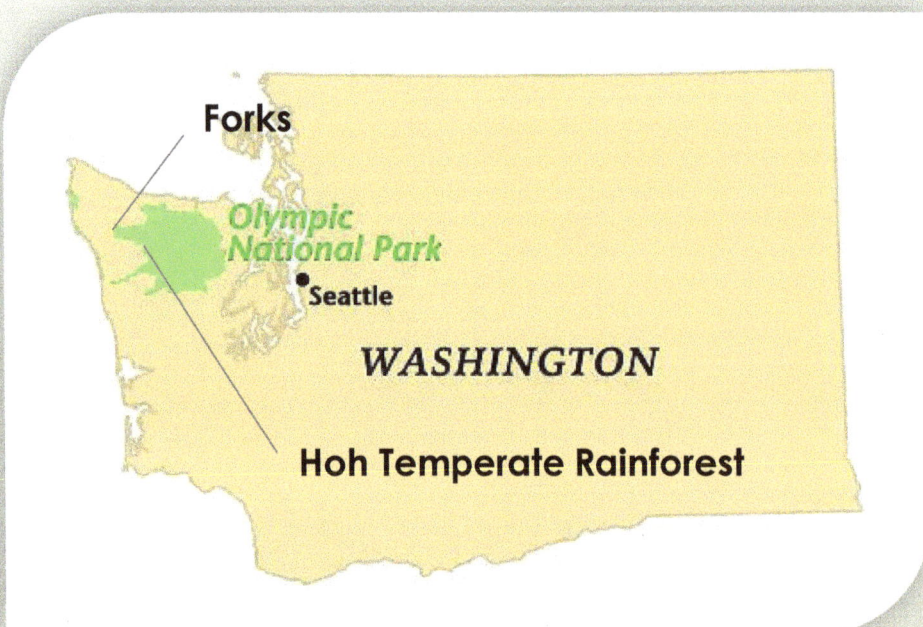

Forks

Olympic
National Park

• Seattle

WASHINGTON

Hoh Temperate Rainforest

diverse

(dī-ˈvərs) 1 differing from one another 2 composed of distinct or unlike elements or qualities

Geography & Climate

The Hoh Rainforest is located in the United States, on the west side of the Olympic Mountains, near Forks, Washington. It averages 140 inches of rainfall per year. Most of this comes from the rain clouds off the Pacific Ocean squeezing out like a sponge as they collide with the mountains. Fog and morning dew also provides moisture. Rushing through the forest is the 50-mile-long Hoh River which begins on the nearby snowy mountain caps. Temperatures here reach a comfortable 65 to 70 degrees in the summer with winter temperatures dropping down to about 40 degrees.

Banana Slug

Fungi

Trail Leading into The Hoh Rainforest

Sword Fern

Lettuce Lichen

Trees & Plants

Hiking through the winding trails, finding chanterelle mushrooms, eating huckleberries, and admiring the sagging moss-covered branches are sights you won't want to miss. Tall trees like the sitka spruce and western hemlock can grow to a remarkable height of over 300 feet tall. The canopy is filled with other trees like the big leaf maple, douglas fir, black cottonwood and the red alder. Under the canopy, you will find plants such as the sword fern, cat-tail moss, and lettuce lichen (ly-kin). These epiphytes cover the trees and forest floor like a blanket. Dogwoods, salal, salmonberries, and Oregon grape are found scattered among the fallen decaying logs. You will also find many varieties of fungi such as bolete (bo-leet), clavaria fungi, and the edible chanterelle mushroom. The combination of rain, mild climate, and a thick carpet of decaying matter provide continual **nourishment** for this diverse plant life.

Salmon

Salmon are born in the fresh waters of the Hoh River and migrate to the salty waters of the Pacific Ocean. They are **anadromous** which means they return to the river to reproduce by laying their eggs. Newly hatched baby salmon are called **alevins**. They live in the fresh waters of the river until they are big enough to swim to the ocean.

Pacific Tree Frog

The pacific tree frog grows up to 2 inches long. It comes in a variety of colors such as green, brown, black, and cream. Sticky toe pads help it to stick to surfaces and climb.

Northern Spotted Owl

Northern spotted owls can grow from 16-19 inches in length. They have a wing span of up to 3 ½ feet. They are dark brown with white spots on their head and breast. These nocturnal birds hunt their prey at night.

Cougar

Cougars are large cats that can weigh up to 150 pounds. They cannot roar like other big cats but rather they purr like a kitten. These cats stay hidden, creeping through the forest until they jump out to pounce on their prey. Baby cougars have spots on their fur to help them stay hidden. They lose them at about 2 months old.

Raccoon

Raccoons are very intelligent. They have amazing dexterity which gives them the ability to open doors, jars, bottles, and latches - making them the bandits of the wild. They make a wide variety of sounds like purrs, snarls, growls, screams, hisses, and whimpers.

Black Bear

Black bears are not always black, they can come in a variety of colors such as silver-blue, gray, and cinnamon-brown. They can weigh between 130 and 500 pounds.

Yellow-pine Chipmunk

Yellow-pine chipmunks grow up to 8 inches in length. They scavenge the forest for fruits, seeds, and a variety of plants like yarrow, thistle, and grass.

Roosevelt Elk

The roosevelt elk is the largest animal in the temperate rainforest and can weigh over 1,000 pounds. This **herbivore** dines on grass, tree bark, and lettuce lichen.

Banana Slug

Banana slugs, as their name implies, are yellow and sometimes have brown spots like ripe bananas. These slimy **gastropods** can grow up to 10 inches long. They have a radula with small teeth. They use this to tear up their food into pieces. If a predator bites off a tentacle and the slug is lucky enough to get away, it has the ability to grow a new one.

Stink Bug

The stink bug has a pair of glands that are filled with a bad smelling fluid. It uses these to defend itself and also to attract a mate.

Olympic Torrent Salamander

Olympic torrent salamanders are dark brown with a yellow belly and grow up to 4 inches in length. They live along the banks of the Hoh River and are occasionally found hiding in damp areas under logs and rocks. These creatures have the ability to grow new limbs if they are unfortunate enough to lose one. They lay their eggs in the river.

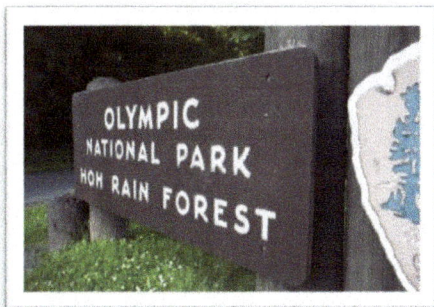

Olympic National Park

As you can see, the Hoh Temperate Rainforest is filled with mysterious life calling you to come explore, see, smell, taste, and experience its beauty for yourself. Filled with leaves made of needles, tart orange salmonberries, and sightings of large purring cats, these forests are truly a wonderland waiting to be discovered.

Chapter

4

Rainforest Plants

Medicine from the Rainforest

Opium Poppy Flower

The seed pod of this flower, not the seeds, contains a drug called morphine. It is used as a strong pain reliever. Doctors will give this to patients after they have had surgery or had an injury. The seeds of this plant are what you are eating when you have a poppy seed muffin.

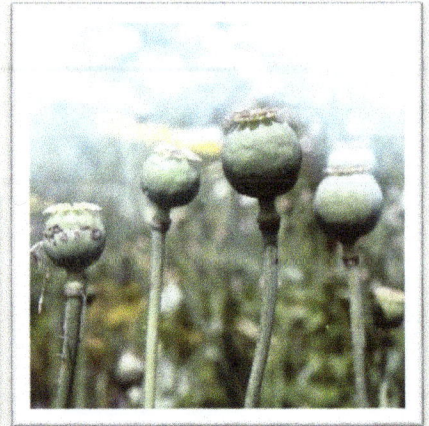

Cinchona Tree

The bark of the cinchona tree contains a drug called quinine. It is used to treat and prevent a deadly disease called malaria. Malaria spreads only by female mosquitos who are infected with it. It is transferred from their saliva into the blood of a human or animal when they bite. Female mosquitos need this blood to produce their eggs. Males feed only on flower nectar.

Madagascar Periwinkle

Two drugs called vincristine and vinblastine are made from Madagascar periwinkle. These are used to treat cancer, Hodgkin's disease and leukemia. The FDA has approved these drugs to be used in chemotherapy. They have increased the survival rate in children with leukemia from 10% to 90%. Leukemia is a type of blood cancer that begins in the bone marrow and can cause a child to become very sick.

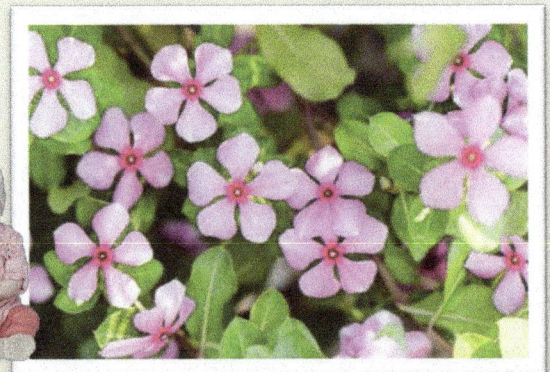

Food from the Rainforest

Pepper

Pepper is the fruit of a plant called Piperaceae. It grows on a vine and is native to India & Indonesia. Peppercorns come in a variety of colors. Black pepper is unripe fruit that has been dried and cooked. Green pepper is treated with a preservative called sulfur dioxide to keep the bright green color. White pepper is made when the outer skin is removed through a process called retting. This means the berries are soaked in water which makes the skin fall off. Pink peppercorns are fully ripened berries.

Dole photo by Copyright: <a href='https://www.123rf.com/profile_carlosyudica'>carlosyudica / 123RF Stock Photo</a>

Pineapple

Nothing compares to the yummy sweetness of a pineapple! This fruit grows on a short stalky stem and takes 2-3 years to ripen. A single plant produces only one pineapple. Once it is harvested the plant dies. You can grow your own by planting the pineapple top in a pot of dirt. If you're patient it will produce a small pineapple. In the early 1900's a man named James Dole moved to Hawaii and planted a pineapple **plantation**. His company is now the leading supplier of pineapples.

plantation (plan-ˈtā-shən) a large area of land where crops are grown and harvested

Peanuts

Peanuts are a legume (leg-oom). Legumes are plants in the pea family. These have pods that grow seeds inside them. Peanuts grow underground like potatoes. It takes 4-5 months until they mature and are ready to harvest. Each plant produces about 40 peanuts. Two United States Presidents, Thomas Jefferson and Jimmy Carter, were peanut farmers. A company called Planters began in 1906. It sells peanut products all over the United States. In 1916, Planters held a contest where people could draw and design a logo for them to use. A 13-year-old boy named Antonio Gentile won! Planters has been using his drawing of Mr. Peanut ever since.

Planters photo Copyright: <a href='https://www.123rf.com/profile_dcwcreations'>dcwcreations / 123RF Stock Photo</a>

Cinnamon

Cinnamon comes from the inner bark of a tree. The two main varieties are the Ceylon Cinnamon Tree and the Cassia Cinnamon Tree. The Cassia variety is cheaper to buy but it does not have the same health benefits as the Ceylon Cinnamon. A good way to tell the difference between the two is by breaking a cinnamon stick in half. If it breaks its Ceylon Cinnamon. Chewing on a cinnamon stick makes an excellent breath freshener.

Cinnamon is in the Bible! *Exodus 30:22-25 says "Take the following fine spices: 500 shekels of liquid myrrh, half as much (that is, 250 shekels) of fragrant cinnamon, 250 shekels of fragrant cane, 500 shekels of cassia- all according to the sanctuary shekel- and a hin of oil. Make these into a sacred anointing oil."*

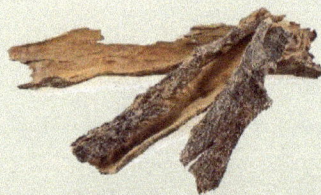

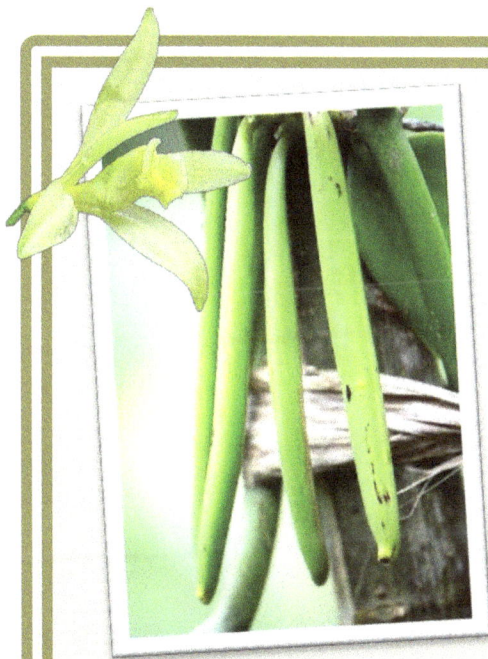

Vanilla

Vanilla is the number one ice-cream flavor in America! It comes from a plant called the Vanilla Bean Orchid which is a type of epiphyte. This trailing vine twists around trees climbing to a height of 300 feet. The flowers have no fragrance and are a light yellow-green color. The fruit is long and skinny like a green bean. The beans take 8-9 months to fully ripen. When you slit open a vanilla bean pod you will see thousands of tiny seeds inside. Fresh pods have no flavor or smell. They must be steamed and fermented in order to release their delicious aroma and taste. Spiders don't like the smell of this fragrant pod which is why it makes a good bug repellent.

This plant is very particular when it comes to pollination. The blooms only open for a few hours before shriveling up. This is why farmers have to hand pollinate every single one making this a very difficult crop to grow. The Melipona bee is the only insect that can pollinate this plant in its native environment. This is because they are the only bee that can fit inside the flower, all other bees are too big.

Sugar Cane

Sugar comes from a type of grass. Its tall stalks can reach up to 30 feet high. It grows only in warm tropical climates. The stalks contain sugar juice which is pressed out. The molasses is separated from the sugar and then it is dried, this makes the white sugar you use. If you want brown sugar the molasses is left in. The remains of the stalk can be burned to generate heat and electricity. It can also be made into paper.

What is an Epiphyte?

Epiphyte (e-pə-ˌfīt)

Did you know there are more than 40,000 different kinds of plants in the rainforest? Warm weather and lots of rain make this the perfect environment for a diverse plant life. We are going to learn about one category of plants called epiphytes. An epiphyte is a plant that grows on another plant. They don't take any nutrients or water from their host, they just use them for support. Wind and birds carry their seeds to nearby trees and branches helping them spread throughout the forest. So how do these plants get the nutrients and water they need? Bromeliads have a cone shaped center. This is where rain, debris, and insects get trapped giving the plant the nutrients it needs. Other epiphytes, such as orchids and ferns, feed themselves by sending out roots to catch the moisture in the air.

Varieties

Some common varieties you would find in a tropical rainforest are bromeliads (brō-ˈmē-lē-ˌad), orchids, (oˈr-kəd) and ferns.

Ferns

Orchids

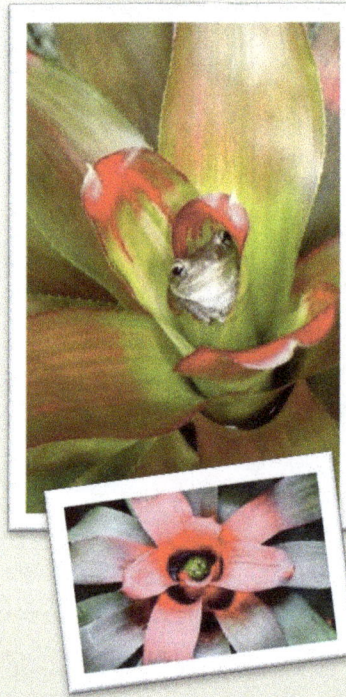

Bromeliad

The poison dart frog lays its eggs in the water-filled center of bromeliads. They use the tiny pools of water to provide a **habitat** for their young high in the trees. When the tadpoles hatch, the parents transport some of them on their backs to other bromeliads so they can have room to grow. Bromeliads provide the necessary home for these **amphibians**.

Liana

Lianas are another plant species you will see twisted around the trees in a rainforest. Though it may look like this plant is growing on another plant, it is not an epiphyte. This plant begins its life on the forest floor rooting itself in the ground. Lianas twist and climb around plants. These vines create a tangled mess while reaching for the sun. With their rope-like vines these plants can grow to an amazing length of 3,000 feet. This creates a forest playhouse for monkeys to swing on. Oftentimes these plants become so entangled that when one tree falls it takes many others with it.

Botany: Giant Water Lily

Leaves

This plant has large waxy-coated leaves that have a rubbery texture. It can grow up to 7 feet wide. The underside is a bright purple color that is covered with sharp needle-like spines with a long stem that reaches to the bottom where it is rooted.

Flower & Seeds

This striking flower can grow up to 18 inches wide. The flower is white at first. After it has been pollinated it turns a beautiful pink color. It has a very delightful sweet smell like a pineapple which attracts insects. The insects pollinate the plant so it can reproduce by growing seeds. It only blooms for 3 days then its drawn into the water where the seeds develop.

Underside view of the leaf and stem.

Scientific Illustration 1.1

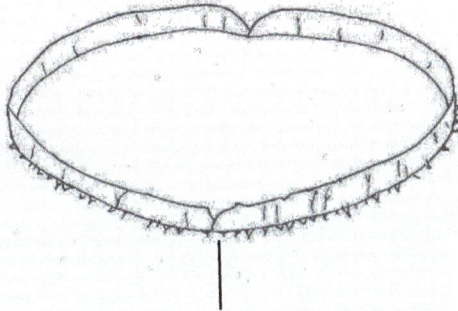

Leaf

Flower- the seeds develop inside

Botany: Titan Arum

Leaves & Flower

This is the smelliest flower in the world and not in the good way. The center of the flower heats up and releases a foul smell like that of rotting flesh. The aroma attracts insects which help pollinate it. Standing 10 feet tall, this striking plant is admired by many people, regardless of the odor, when the flower is in bloom. It doesn't always have a flower though. It booms only once every four to five years. The remainder of the time it grows a tall spike covered in leaves. It is native to the rainforests in Sumatra.

Scientific Illustration 1.2

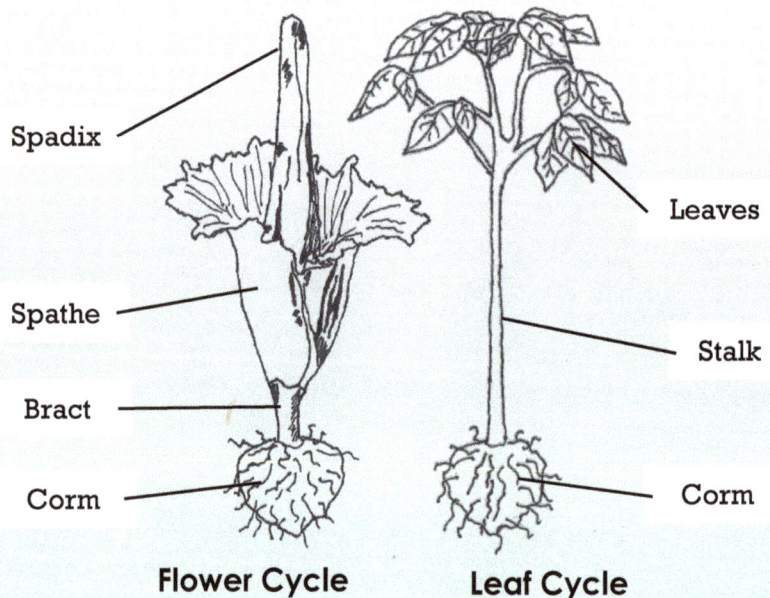

Spadix

Spathe

Bract

Corm

Leaves

Stalk

Corm

Flower Cycle   **Leaf Cycle**

Botany: Passion Flower

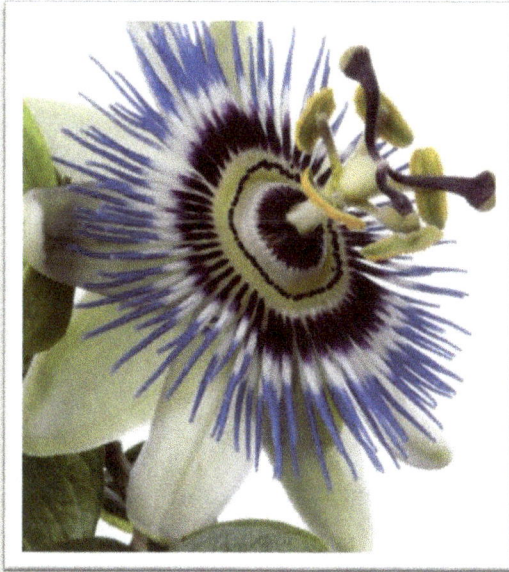

The passion flower, also called passiflora, is a vine that spirals and twists up to 30 feet in length. Standing out against the green leaves are the most elegant eye-catching flowers. These flowers range in size and color. Most species display 3 to 4-inch blooms. It is native to South America. Hummingbirds, bees, bats, and butterflies dine on its sweet nectar.

Fruit

This plant produces an exotic tropical fruit called passion fruit. Depending on the species, the fruit can be anywhere from yellow-orange to plum-purple in color. Some people like to eat it fresh but it is most often made into juice.

Leaves and Roots

This plant has a long history of being used as medicine for a variety of conditions. Most commonly it is used in tea to help people fall asleep.

Scientific Illustration 1.3

Flower

Fruit

Stigma

Leaf

Botany: Rafflesia

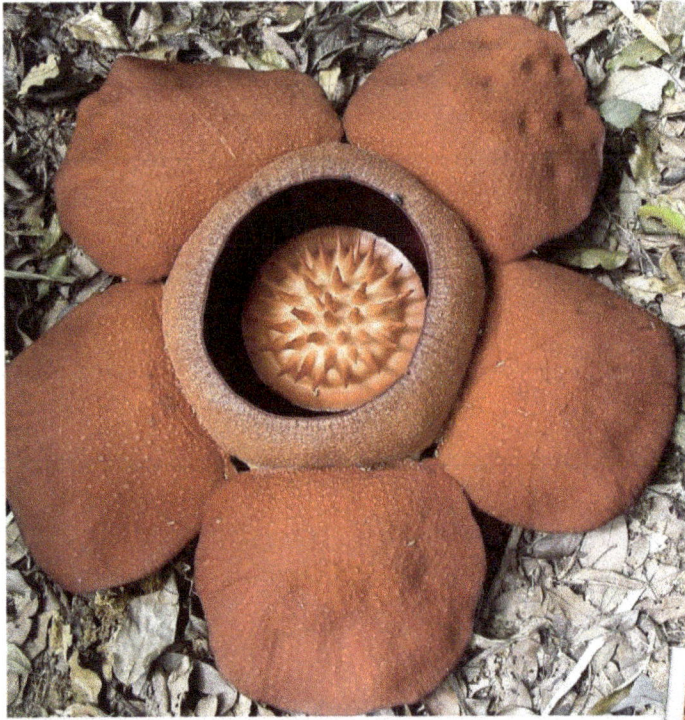

(rə-ˈflē-zh(ē-)ə)
This plant is native to Asian rainforests. It has no stems, leaves, or true roots. Being a **parasite**, it can only live by inserting its nutrient-sucking organ inside a vine. This organ can be thought of as its root system where is draws up food and water from its host.

parasite (per-ə-ˌsīt) an animal or plant that lives in or on another animal or plant and gets food or protection from it

Flower

The flower can grow up to 3 feet wide and may weigh as much as 25 pounds. Like the titan arum, this plant has a foul odor that smells like rotting meat. The bud takes months to develop. It only blooms for 3 days making it extremely difficult to locate in the rainforest.

Scientific Illustration 1.4

Perigone Lobe

Diaphragm

Anther

Vine Host

Chapter

5

Rainforest Animals

Hummingbird

Hummingbirds are named after the humming sound their wings make as they fly around from flower to flower. There are more than 300 varieties of this species and they all live in North and South America. The miniature Bee Hummingbird is the smallest one. It grows to a miniscule size of only 2 inches long. The largest of the species is the Giant Hummingbird from South America. It can grow up to 8 inches long.

Hummingbirds use their long tongues to lap up nectar found in the center of flowers. This species of bird has a greater advantage over other birds in finding food because they can hover over plants to eat. Other birds must perch on a branch making it harder to reach certain kinds of food.

Migration

Hummingbirds begin their migration to Mexico and Central America in July. In order to make this long nonstop flight, of up to 500 miles, they will store up a layer of fat that is equal to half their weight.

Pollination

Some plants grow two different kinds of flowers. These are male and female. In order for the flower to reproduce, the pollen from the male flower must be transferred to the center of the female flower. A hummingbird is the perfect pollinator because it will transfer the pollen as it visits each flower, eating the sweet nectar. The hummingbirds and these flowering plants have a **symbiotic** relationship. This means they each need the other for survival.

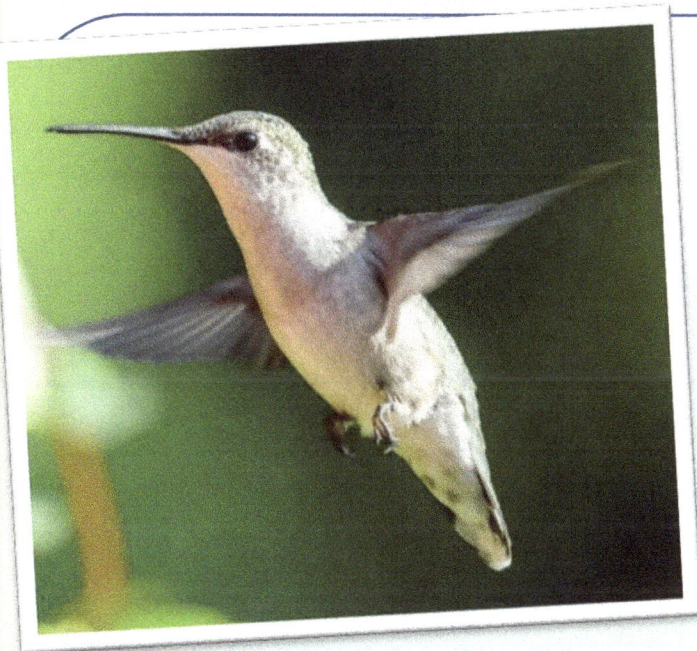

Unique Features

These tiny birds can't walk or jump because their legs are too short. If they were to injure their wings they wouldn't be able to move and their chance for survival would be slim. A long tongue makes it easy for them to reach the nectar in the center of a flower.

Ecosystem

God gave hummingbirds a special purpose making them important to our ecosystem. An ecosystem is the cycle of life working together. This is when plants, animals, organisms, water, and soil support each other for continual survival. Without hummingbirds, many plants and animals would become extinct.

Is That a Hummingbird?

There is a moth that looks and behaves like the hummingbird. It's called the Hummingbird Hawk-Moth. It's just a little smaller than the Bee Hummingbird and it feeds on the same flowers using a proboscis.

Sword-billed

Broad Billed

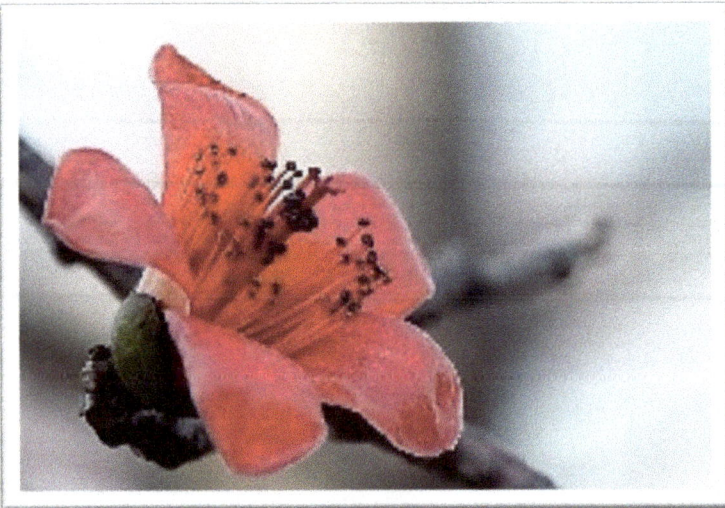

Flowers

Growing plants that have bright colored flowers will attract hummingbirds. In the spring time, you can create a hummingbird garden by planting their favorite flowers such as cardinals, bee balm, columbine, and zinnias. Other flowers that will attract these birds are lupine, salvia, bleeding hearts, trumpet creeper and petunias.

Predators

Cats, snakes, and hawks are common **predators** of the hummingbird. There are a few animals that have been known to hunt and kill hummingbirds that you may not think of. These are the praying mantis, spiders, wasps, and even frogs! Eyelash vipers are also predators of the hummingbird. They have scales above their eyes that look like eyelashes. These vipers have been known to hide near flowers and snatch a hummingbird for dinner.

World's Strongest Bird   Did you know that...

- Hummingbirds can fly upside down, forward, and backward and can hover like a helicopter?
- They can take 250 breaths per minute?
- They can lick 10-15 times a second?
- They can fly 30-60 miles per hour and fly 500 miles nonstop across the Gulf of Mexico during migration? (How long would that take them at 30 miles per hour?)
- They are fearless and aggressive when protecting their nest... and don't even think about messing with their flowers; they are very territorial.
- Their large pectoral muscles make up 30% of their weight to help those wings move with speed?
- Their wings beat a whopping 50-200 times per second?

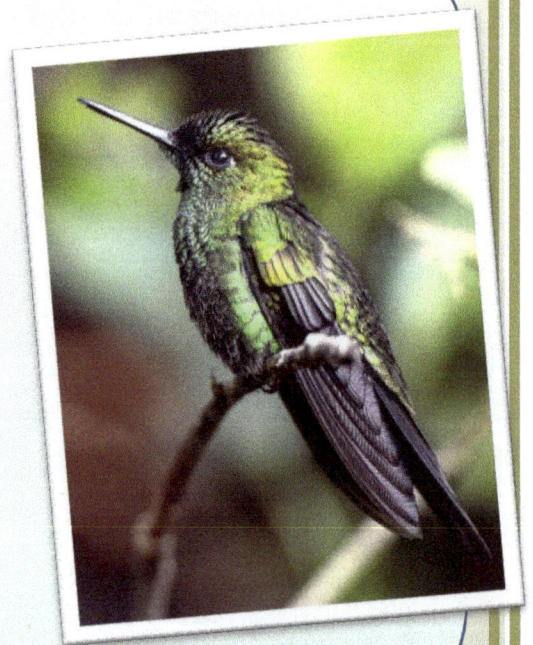

Wing Structure

Hummingbirds have a wing structure different from all other birds in the world. Their shoulder has a ball joint which allows them to rotate it 180 degrees. This gives them the ability to hover and even fly upside down. Their wing bones are very porous and hollow which makes them light weight. The pattern of movement their wings make when flying is in an oval. When they hover, their wings move in a figure-8 pattern.

Violet Saber Wing

Bird Feeder

Attract hummingbirds by setting out a bird feeder. You can make your own sweet nectar for them by mixing 1 cup of warm water with ¼ cup of sugar, stir until the sugar is dissolved. Next, pour it into a hummingbird feeder. Be sure to hang it at least 5 feet off the ground to protect them from those kitty cats!

Ruby-throated Hummingbird

The ruby-throated hummingbird is a common species in North America. It measures 3 ½ inches in length. It lays 1-2 eggs in a nest that measures slightly over 2 inches. The eggs are about the size of a jellybean- only ¼ to ½ inch long. When it builds a nest, it uses spider web silk to attach it to a branch. The eggs take 16-18 days to fully mature. When the baby birds hatch they are only 1 inch in length. They remain in the nest until they are fully grown. The smallest feathers in world cover their bodies.

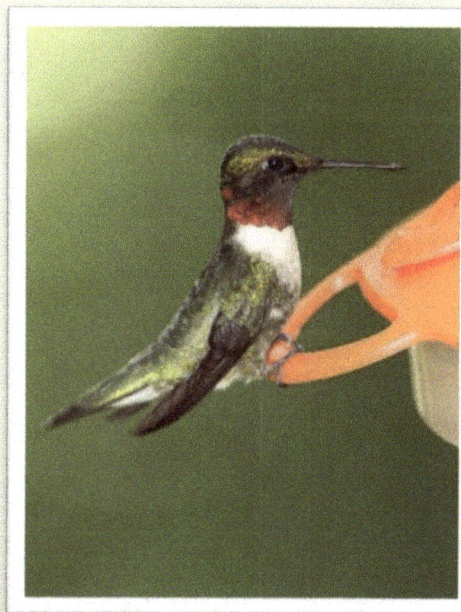

Animal Classification

What is Animal Classification?

Scientists classify living organisms into groups to show similarities and differences. This helps us to understand them better. There are a lot of different ways all living things are grouped. One group is called the animal kingdom. There are 2 main groups of animals: invertebrates and vertebrates. These are then divided into smaller sub-groups or species, such as sponges, spiders, fish, and mammals.

Invertebrate

Invertebrates are animals that have no backbone and are **cold-blooded**. Many of them have a hard shell like a crab. Others, like the worm or the jellyfish, have a **hydrostskeleton**. This is a water-based skeleton inside their body where their organs float in fluid for protection. More than ninety percent of all animals fall into this group. There are about 1,300,000 species of invertebrates in the world!

Vertebrate

Vertebrates have backbones. This group is much smaller with about 64,000 species. The animals in this group are generally much larger in size. Because they have spinal cords, nervous systems, and brains, these animals are very smart. This makes them seem like Einsteins compared to invertebrates. The inner skeleton and muscles give them a greater ability to move as well.

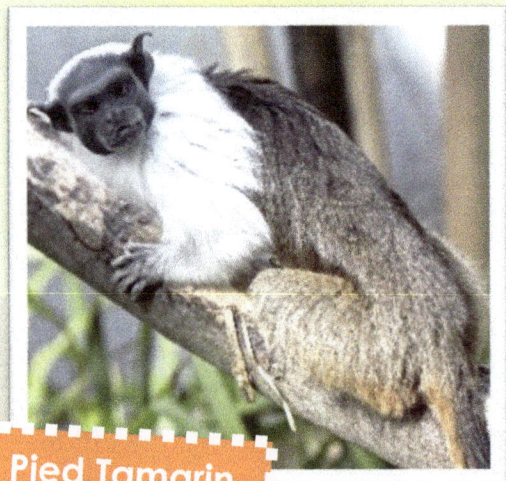

Pied Tamarin

Invertebrate Animals

Crab
Crabs are a type of crustacean. They have five pairs of legs. Two of them have pinchers.

Coral
Coral lives in colonies in the ocean. They have a hard-external skeleton.

Sponge
Sponges are an animal with a soft porous body that lives in the ocean.

Spider
Spiders are arachnids. They have eight legs and fangs which they use to inject poison into their prey.

Insect
Insects belong to the family of arthropods. They have six legs and some have wings.

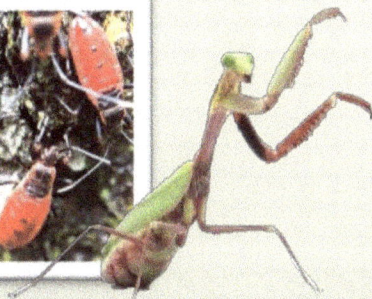

Worm
Worms are a burrowing animal. They have a soft body and no limbs.

Vertebrate Animals

Amphibian
Amphibians have babies that breathe through gills then mature into lung breathing adults.

Fish
Fish are a limbless animal that lives in the water. They swim with fins and breathe with gills.

Bird
Birds lays eggs. They fly and have feathers and wings. They and are **warm-blooded**.

Reptile
Reptiles are cold-blooded. They lay soft shelled eggs and have dry scaly skin.

Mammal
Mammals are warm-blooded and have hair or fur. They give birth to their young and feed them with milk.

Food Chain

Decomposers break down dead or rotting material which gives nutrients to plants through the soil.

When **Meat Consumers** die they are broken down by decomposers returning the nutrients back into the soil. This begins the cycle all over again.

Producers are plants that provide nutrients for plant eating animals.

Plant Consumers are eaten by meat consumers giving them the nutrients they need.

Glossary

A glossary is a list that provides definitions for the difficult or unusual words used in a book.

1. **alevin** (a-lə-vən) a young fish; *especially*: a newly hatched salmon
2. **amphibian** (am-ˈfi-bē-ən) an animal (such as a frog) that can live both on land and in water
3. **anadromous** (ə-ˈna-drə-məs) migrating from salt water to spawn in fresh water
4. **bacteria** (bak-ˈtir-ē-ə) a group of very small living things that often cause disease
5. **biome** (bī-ˌōm) a community of distinctive plants and animals living together in a particular climate and physical environment
6. **bromeliad** (brō-ˈmē-lē-ˌad) a plant typically having short stems with rosettes of stiff, usually spiny, leaves
7. **camouflage** (ka-mə-ˌfläj) something (such as color or shape) that protects an animal from attack by making the animal difficult to see in the area around it
8. *carnivore* (kär-nə-ˌvȯr) an animal that feeds on meat
9. **climate** (klaɪmət) a region with particular weather patterns or conditions
10. **cold-blooded** (kōld-ˈblə-dəd) having a body temperature that varies with the temperature of the environment
11. **debris** (də-ˈbrē) pieces left from something broken down or destroyed
12. **deciduous** (di-ˈsi-jə-wəs) leaves that fall off every year
13. **decompose** (decomposition) (dē-kəm-ˈpōz) to break down or be broken down into simpler parts or substances especially by the action of living things (as bacteria or fungi)
14. **diverse** (dī-ˈvərs) different from each other
15. **dormant** (dȯr-mənt) not actively growing but protected from the environment
16. **ecosystem** (ē-kō-ˌsi-stəm) the whole group of living and nonliving things that make up an environment and affect each other
17. **endemic** (en-ˈde-mik) growing or existing in a certain place or region
18. **epiphyte** (e-pə-ˌfīt) a plant that derives its moisture and nutrients from the air and rain and grows usually on another plant
19. **equator** (i-ˈkwā-tər) an imaginary circle around the earth everywhere equally distant from the north pole and south pole
20. **evergreen** (e-vər-ˌgrēn) having leaves that remain green all year long
21. **fungi** (fəŋ-ˌgī) living things such as mushrooms and molds that have no chlorophyll, must live in or on plants, animals, or decaying material
22. **gastropod** (ga-strə-ˌpäd) any mollusk of the class Gastropoda, comprising the snails, whelks, slugs, etc.
23. **geology** (jē-ˈä-lə-jē) the rocks, land, process of land formation, etc., of a particular area

24. **gravity** (ˈgra-və-tē) the natural force that tends to cause physical things to move towards each other: the force that causes things to fall towards the Earth
25. **habitat** (ha-bə-ˌtat) the place or environment where a plant or animal naturally or normally lives and grows
26. **herbivore** (hər-bə-ˌvȯr) an animal that only eats plants
27. **hibernate** (hī-bər-ˌnāt) to pass all or part of winter in an inactive state in which the body temperature drops and breathing slows
28. **humid** (hyü-məd) containing moisture
29. **hydroskeleton** (hyˈdro·ske-lə-tən) a structure found in many soft-bodied animals, consisting of a fluid-filled cavity
30. **indigenous** (in-ˈdij-ə-nəs) produced, living, or existing naturally in a particular region or environment
31. **jungle** (jəŋ-gəl) a large area of land covered with a thick tangled growth of plants
32. **latex** (lā-ˌteks) a milky plant juice that is the source of rubber
33. **mammal** (ma-məl) a type of animal that feeds milk to its young and that usually has hair or fur covering most of its skin
34. **migrate** (mī-ˌgrāt) to pass, usually periodically, from one region or climate to another for feeding or breeding
35. **native** (nā-tiv) living or growing naturally in a certain region
36. **nectar** (nek-tər) a sweet liquid produced by plants
37. **nocturnal** (näk-ˈtər-nᵊl) active at night
38. **nourishment** (nər-ish-mənt) food and other things that are needed for health, growth, etc.
39. **omnivore** (äm-ni-ˌvȯr) an animal that eats both plants and other animals
40. **parasite** (per-ə-ˌsīt) an animal or plant that lives in or on another animal or plant and gets food or protection from it
41. **plantation** (plan-ˈtā-shən) a large area of land where crops are grown and harvested
42. **precipitation** (pri-ˌsi-pə-ˈtā-shən) water that falls to the earth as hail, mist, rain, sleet, or snow
43. **predator** (pre-də-tər) an animal that lives mostly by killing and eating other animals
44. **prehensile** (prē-ˈhen(t)-səl) capable of grabbing or holding something by wrapping around it
45. **prey** (prā) an animal that is hunted or killed by another animal for food
46. **proboscis** (prə-ˈbäs-əs, -kəs) the long, thin nose of some animals
47. **species** (spē-shēz) a group of animals or plants that are similar
48. **symbiotic** (symˈbi·otˈic) the relationship between two different kinds of living things that live together and depend on each other
49. **warm-blooded** (wȯrm-ˈblə-dəd) having a body temperature that does not change when the temperature of the environment changes

Political Map of the World

Physical Map of the World

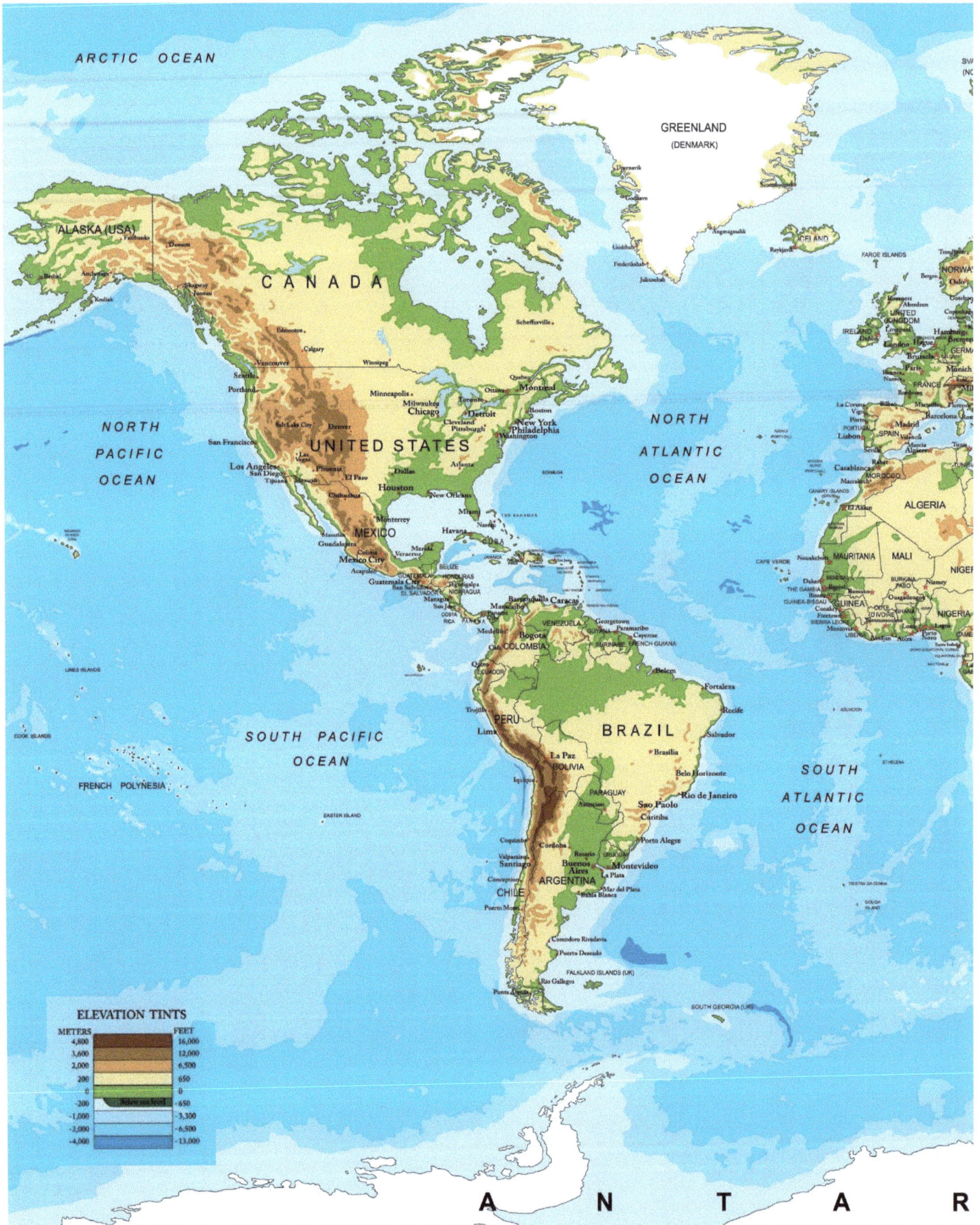

ARCTIC OCEAN

GREENLAND
(DENMARK)

ICELAND

FAROE ISLANDS

NORWAY

ALASKA (USA)

C A N A D A

Edmonton

Calgary

Vancouver
Seattle
Portland

Winnipeg

Minneapolis
Chicago
Milwaukee

Quebec
Montreal
Ottawa
Toronto
Detroit
Cleveland
Pittsburgh

Boston
New York
Philadelphia
Washington

Scheffervile

IRELAND
UNITED
KINGDOM
Dublin

Aberdeen

London
Brussels
Paris

Hamburg
GERMANY
Munich

NORTH
PACIFIC
OCEAN

San Francisco

Salt Lake City
Denver

U N I T E D   S T A T E S

Las
Vegas
Los Angeles
San Diego
Tijuana

Phoenix
El Paso

Dallas
Houston

Atlanta

New Orleans

Miami

NORTH
ATLANTIC
OCEAN

PORTUGAL
Lisbon

SPAIN
Madrid
Barcelona

Valencia
Seville

MOROCCO
Casablanca
Marrakesh

ALGERIA

Chihuahua

Monterrey

MEXICO

Guadalajara
Colima
Mexico City
Acapulco

Merida
Veracruz

Havana

CUBA

THE BAHAMAS
Nassau

CANARY ISLANDS

CAPE VERDE

MAURITANIA

MALI

NIGER

GUATEMALA
HONDURAS
Guatemala City
El Salvador
NICARAGUA

BELIZE

Managua

COSTA
RICA
PANAMA

Barranquilla
Maracaibo

Caracas

VENEZUELA

Georgetown
GUYANA
Paramaribo
Cayenne
SURINAME FRENCH GUIANA

THE GAMBIA
GUINEA-BISSAU
Conakry
SIERRA LEONE

BURKINA
FASO
Ouagadougou

GUINEA

CÔTE
D'IVOIRE

Niamey

NIGERIA

LIBERIA

Medellin

Bogota
COLOMBIA

Quito
ECUADOR

Belem

Fortaleza

Trujillo
PERU
Lima

La Paz
BOLIVIA

Iquique

B R A Z I L

Brasília

Recife

Salvador

Belo Horizonte

SOUTH
ATLANTIC
OCEAN

SOUTH PACIFIC
OCEAN

FRENCH POLYNESIA

EASTER ISLAND

COOK ISLANDS

LINE ISLANDS

PARAGUAY
Asuncion

Sao Paolo
Curitiba

Rio de Janeiro

Porto Alegre

Coquimbo

Valparaiso
Santiago
Concepcion

Cordoba
Rosario
URUGUAY
Buenos
Aires
La Plata
Mar del Plata
Bahia Blanca

Montevideo

ST HELENA

TRISTAN DA CUNHA

CHILE

A R G E N T I N A

Puerto Montt

Comodoro Rivadavia

Puerto Desado

FALKLAND ISLANDS (UK)
Rio Gallegos
Punta Arenas

SOUTH GEORGIA (UK)

ELEVATION TINTS

METERS		FEET
4,800		16,000
3,600		12,000
2,000		6,500
200		650
0		0
-200	Below sea level	-650
-1,000		-3,300
-2,000		-6,500
-4,000		-13,000

A N T A R

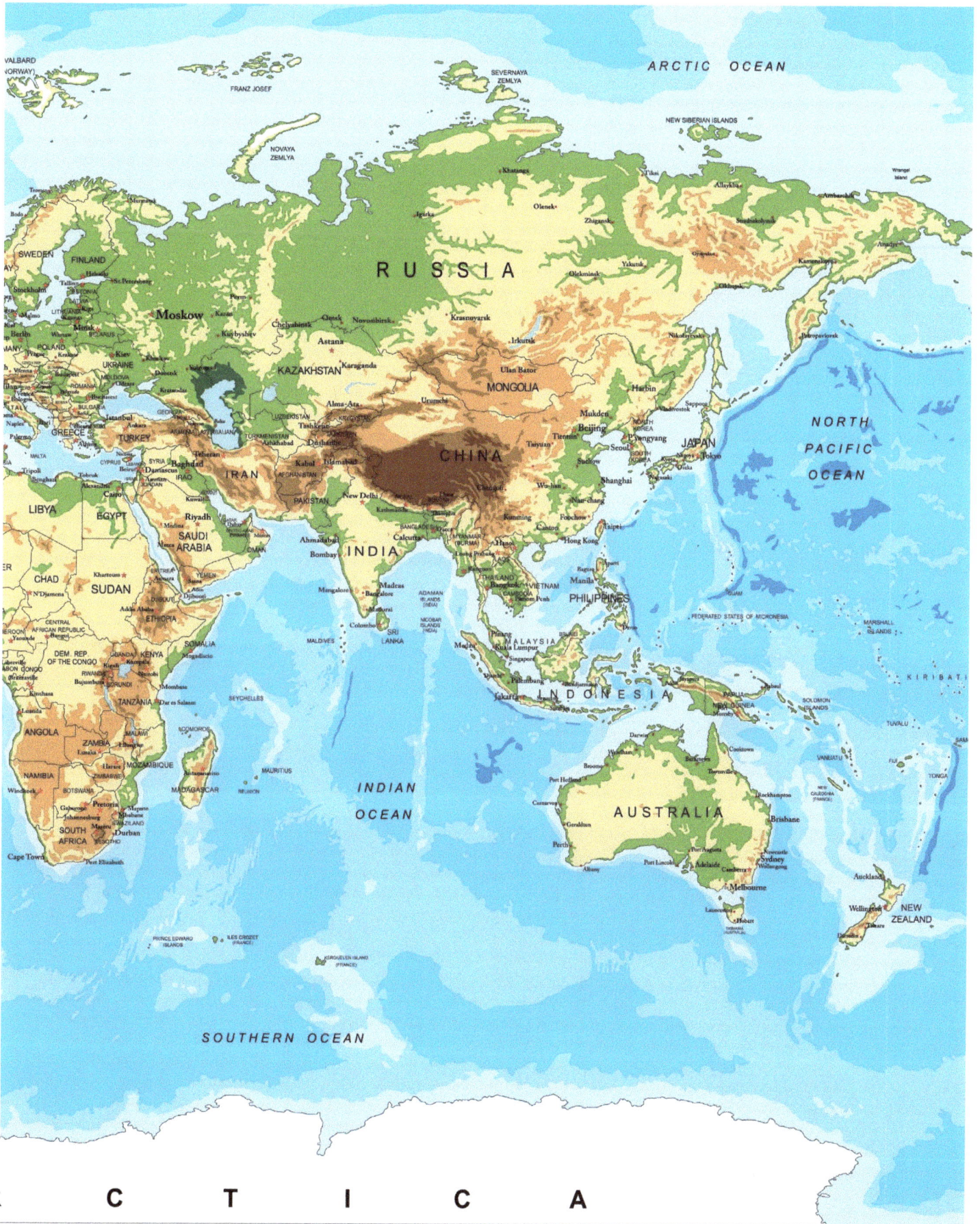

ARCTIC OCEAN

SVALBARD
(NORWAY)
FRANZ JOSEF
SEVERNAYA
ZEMLYA
NEW SIBERIAN ISLANDS
Wrangel
Island
NOVAYA
ZEMLYA
Khatanga
Tiksi
Allykhya
Ambarchik
Olenek
Zhigansk
Anadyr
Igarka
SWEDEN
FINLAND
Yakutsk
Kamenskoya
Murmansk
St. Petersburg
Olekminsk
RUSSIA
Perm
Stockholm
Tallinn
Srednekolymsk
Moskow
Kazan
Oymakon
Olenek
Novosibirsk
Oyapan
GERMANY
POLAND
Chelyabinsk
Krasnoyarsk
Nikolayevsk
Petropavlovsk
Kuybyshev
Astana
Irkutsk
UKRAINE
KAZAKHSTAN
Karaganda
Ulan Bator
Harbin
MONGOLIA
Sapporo
Vladivostok
GEORGIA
UZBEKISTAN
Urumchi
Mukden
NORTH
KOREA
JAPAN
NORTH
TURKEY
Tashkent
Alma-Ata
Tsitsihar
Pyongyang
Seoul
Tokyo
PACIFIC
GREECE
Teheran
Dushanbe
Tientsin
Beijing
SOUTH KOREA
OCEAN
IRAN
Baghdad
Kabul
CHINA
Tsiyuan
Nagasaki
Damascus
AFGHANISTAN
Islamabad
Shanghai
Wu-han
LIBYA
EGYPT
Riyadh
New Delhi
Kashmir
Chungking
Wan-chang
SAUDI
ARABIA
PAKISTAN
Kunming
Foochow
Taipei
Medina
Canton
Hong Kong
CHAD
SUDAN
Khartoum
YEMEN
Ahmadabad
BANGLADESH
Bombay
INDIA
MYANMAR
(BURMA)
Hanoi
Manila
ETHIOPIA
Madras
THAILAND
VIETNAM
Bangalore
Bangkok
PHILIPPINES
SOMALIA
Mogadishu
ANDAMAN
ISLANDS
(INDIA)
Phnom Penh
FEDERATED STATES OF MICRONESIA
MARSHALL
ISLANDS
KENYA
Nairobi
NICOBAR
ISLANDS
(INDIA)
MALDIVES
Colombo
SRI
LANKA
MALAYSIA
Kuala Lumpur
KIRIBATI
TANZANIA
Dar es Salaam
Singapore
TUVALU
SEYCHELLES
INDONESIA
Jakarta
PAPUA
NEW GUINEA
SOLOMON
ISLANDS
ANGOLA
ZAMBIA
MAURITIUS
NAMIBIA
MOZAMBIQUE
INDIAN
VANUATU
NEW
CALEDONIA
(FRANCE)
MADAGASCAR
OCEAN
Darwin
AUSTRALIA
Brisbane
SOUTH
AFRICA
Cape Town
PRINCE EDWARD
ISLANDS
ILES CROZET
(FRANCE)
Perth
Adelaide
Sydney
Melbourne
Auckland
Wellington
NEW
ZEALAND
KERGUELEN ISLAND
(FRANCE)

SOUTHERN OCEAN

C T I C A

95

Resources Used

WEBSITES USED:
decah.com, pbs.org, interesting-africa-facts.com, rainforest-australia.com, earchwatch.com, blueplanetbiomes.com, panda.org, lifeofplnat.blogspot.com, australia.com, mongabay.com, worldwildlife.org, biology4kids.com, australiangeographic.com.au, nutindustry.org.au, avocadosource.com, wettropics.gov.au, arf.net.au, tasteaustralia.biz, globalforestatlas.yale.edu, planters.com, flowers.about.com, mexicanvanilla.com, time4writing.com, rainforest-alliance.org, kids.ct.gov, photographicdictionary.com, hummingbirds.net, www.merriam-webster.com, www.britannica.com

page 33 Acai Plam phot By Dick Culbert from Gibsons, B.C., Canada - Euterpe oleracea, CC BY 2.0,
https://commons.wikimedia.org/w/index.php?curid=51780227
page 33 Brazil nut fruit photo By Lior Golgher - Own work, GFDL,
https://commons.wikimedia.org/w/index.php?curid=3231680
page 37 Pygmy Hippo photo by By Chuckupd - Transferred from en.wikipedia to Commons by
Berichard using CommonsHelper., Public Domain,
https://commons.wikimedia.org/w/index.php?curid=4574303
page 37 Goliath beetle photo by CC BY-SA 3.0,
https://commons.wikimedia.org/w/index.php?curid=355188
page 38 Wild celery photo by By Kristian Peters -- Fabelfroh 14:38, 27 June 2007 (UTC) -
photographed by Kristian Peters, CC BY-SA 3.0,
https://commons.wikimedia.org/w/index.php?curid=2312351
page 38 Impatiens photo by By Denis Barthel at the German language Wikipedia, CC BY-SA 3.0,
https://commons.wikimedia.org/w/index.php?curid=26265771
page 38 Impatiens photo close up by By Cbaile19 - Own work, CC0,
https://commons.wikimedia.org/w/index.php?curid=44451235
page 42 Madagascar civet photo by By Joaquín Romero Redondo - Own work, CC BY-SA 3.0,
https://commons.wikimedia.org/w/index.php?curid=31994601
page 44 Alluaudia plant photo by By Krzysztof Ziarnek, Kenraiz - Own work, CC BY-SA 4.0,
https://commons.wikimedia.org/w/index.php?curid=57192929
page 57 Big leaf Mahogany photo by By jayeshpatil912 -
http://www.flickr.com/photos/75380256@N06/6925165314/, CC BY 2.0,
https://commons.wikimedia.org/w/index.php?curid=25006619
page 67 Torrent salamander photo by By Hollingsworth John and Karen, U.S. Fish and Wildlife Service
- http://www.public-domain-image.com/public-domain-images-pictures-free-stock-photos/fauna-
animals-public-domain-images-pictures/reptiles-and-amphibians-public-domain-images-
pictures/lizards-and-geckos-pictures/olympic-salamander.jpg, Public Domain,
https://commons.wikimedia.org/w/index.php?curid=24885886
page 76 Titan arum close up phot by By Fbianh - Own work, CC0,
https://commons.wikimedia.org/w/index.php?curid=40806448
page 79 Rafflesia life cycle photo by By Dakuhippo - i drew itPreviously published:
http://archive.org/details/Lifecycles, CC BY-SA 3.0,
https://commons.wikimedia.org/w/index.php?curid=28727106
Page 83 Hummingbird Hawk Moth photo by By By Wikipedia editor J-E Nyström (User:Janke), Finland,
CC BY-SA 2.5, https://commons.wikimedia.org/w/index.php?curid=1158731